HIGH-SPEED DIGITAL SYSTEM DESIGN

HIGH-SPEED DIGITAL SYSTEM DESIGN

A HANDBOOK OF INTERCONNECT THEORY AND DESIGN PRACTICES

Stephen H. Hall
Garrett W. Hall
James A. McCall

A Wiley-Interscience Publication

JOHN WILEY & SONS, INC.

New York · Chichester · Weinheim · Brisbane · Singapore · Toronto

This book is printed on acid-free paper. ∞

Copyright © 2000 by John Wiley & Sons, Inc. All rights reserved.

Published simultaneously in Canada.

For ordering and customer service, call 1-800-CALL-WILEY.

Library of Congress Cataloging-in-Publication Data:

Hall, Stephen H.
    High-speed digital system design: a handbook of interconnect theory
and design practices/Stephen H. Hall, Garrett W. Hall, James A. McCall
        p.   cm.
    ISBN 0-471-36090-2 (cloth)
    1. Electronic digital computers—Design and construction. 2. Very
high speed integrated circuits—Design and construction.
    3. Microcomputers—Buses. 4. Computer interfaces. I. Hall, Garrett W.
    II. McCall, James A.   III. Title.

TK7888.3 H315   2000
621.39′8—dc21   00-025717

Printed in the United States of America

10 9 8 7 6 5 4 3 2

CONTENTS

◼◼◼◼ PREFACE

This book covers the practical and theoretical aspects necessary to design modern high-speed digital systems at the platform level. The book walks the reader through every required concept, from basic transmission line theory to digital timing analysis, high-speed measurement techniques, as well as many other topics. In doing so, a unique balance between theory and practical applications is achieved that will allow the reader not only to understand the nature of the problem, but also provide practical guidance to the solution. The level of theoretical understanding is such that the reader will be equipped to see beyond the immediate practical application and solve problems not contained within these pages. Much of the information in this book has not been needed in past digital designs but is absolutely necessary today. Most of the information covered here is not covered in standard college curricula, at least not in its focus on digital design, which is arguably one of the most significant industries in electrical engineering.

The focus of this book is on the design of robust high-volume, high-speed digital products such as computer systems, with particular attention paid to computer busses. However, the theory presented is applicable to any high-speed digital system. All of the techniques covered in this book have been applied in industry to actual digital products that have been successfully produced and sold in high volume.

Practicing engineers and graduate and undergraduate students who have completed basic electromagnetic or microwave design classes are equipped to fully comprehend the theory presented in this book. At a practical level, however, basic circuit theory is all the background required to apply the formulas in this book.

Chapter 1 describes why it is important to comprehend the lessons taught in this book. (Authored by Garrett Hall)

Chapter 2 introduces basic transmission line theory and terminology with specific digital focus. This chapter forms the basis of much of the material that follow. (Authored by Stephen Hall)

Chapters 3 and 4 introduce crosstalk effects, explain their relevance to digital timings, and explore nonideal transmission line effects. (Authored by Stephen Hall)

Chapter 5 explains the impact of chip packages, vias, connectors, and many other aspects that affect the performance of a digital system. (Authored by Stephen Hall)

Chapter 6 explains elusive effects such as simultaneous switching noise and nonideal current return path distortions that can devastate a digital design if not properly accounted for. (Authored by Stephen Hall)

Chapter 7 discusses different methods that can be used to model the output buffers that are used to drive digital signals onto a bus. (Authored by Garrett Hall)

Chapter 8 explains in detail several methods of system level digital timing. It describes the theory behind different timing schemes and relates them to the high-speed digital effects described throughout the book. (Authored by Stephen Hall)

Chapter 9 addresses one of the most far-reaching challenges that is likely to be encountered: handling the very large number of variables affecting a system and reducing them to a manageable methodology. This chapter explains how to make an intractable problem tractable. It introduces a specific design methodology that has been used to produce very high performance digital products. (Authored by Stephen Hall)

Chapter 10 covers the subject of radiated emissions, which causes great fear in the hearts of system designers because radiated emission problems usually cannot be addressed until a prototype has been built, at which time changes can be very costly and time-constrained. (Authored by Garrett Hall)

Chapter 11 covers the practical aspects of making precision measurements in high-speed digital systems. (Authored by James McCall)

Acknowledgments

Many people have contributed directly or indirectly to this book. We have been fortunate to keep the company of excellent engineers and fine peers. Among the direct, knowing contributors to this book are:

Dr. Maynard Falconer, Intel Corporation

Mike Degerstrom, Mayo Foundation, Special Purpose Processor Development Group

Dr. Jason Mix, Intel Corporation

Dorothy Hall, PHI Incorporated

We would also like to recognize the following people for their continuing collaboration over the years, which have undoubtedly affected the outcome of this book. They have our thanks.

Howard Heck, Intel Corporation; Oregon Graduate Institute

Michael Leddige, Intel Corporation

Dr. Tim Schreyer, Intel Corporation

Harry Skinner, Intel Corporation

Alex Levin, Intel Corporation

Rich Melitz, Intel Corporation

Wayne Walters, Mayo Foundation, Special Purpose Processor Development Group

Pat Zabinski, Mayo Foundation, Special Purpose Processor Development Group

Dr. Barry Gilbert, Mayo Foundation, Special Purpose Processor Development Group

Dr. Melinda Picket-May, Colorado State University

Special thanks are also given to Jodi Hall, Stephen's wife, without whose patience and support this book would not have been possible.

CHAPTER 1

The Importance of Interconnect Design

The speed of light is just too slow. Commonplace, modern, volume-manufactured digital designs require control of timings down to the picosecond range. The amount of time it takes light from your nose to reach your eye is about 100 picoseconds (in 100 ps, light travels about 1.2 in.). This level of timing must not only be maintained at the silicon level, but also at the physically much larger level of the system board, such as a computer motherboard. These systems operate at high frequencies at which conductors no longer behave as simple wires, but instead exhibit high-frequency effects and behave as transmission lines that are used to transmit or receive electrical signals to or from neighboring components. If these transmission lines are not handled properly, they can unintentionally ruin system timing. Digital design has acquired the complexity of the analog world and more. However, it has not always been this way. Digital technology is a remarkable story of technological evolution. It is a continuing story of paradigm shifts, industrial revolution, and rapid change that is unparalleled. Indeed, it is a common creed in marketing departments of technology companies that "by the time a market survey tells you the public wants something, it is already too late."

This rapid progress has created a roadblock to technological progress that this book will help solve. The problem is that modern digital designs require knowledge that has formerly not been needed. Because of this, many currently employed digital system designers do not have the knowledge required for modern high-speed designs. This fact leads to a surprisingly large amount of misinformation to propagate through engineering circles. Often, the concepts of high-speed design are perceived with a sort of mysticism. However, this problem has not come about because the required knowledge is unapproachable. In fact, many of the same concepts have been used for several decades in other disciplines of electrical engineering, such as radio-frequency design and microwave design. The problem is that most references on the necessary subjects are either too abstract to be immediately applicable to the digital designer, or they are too practical in nature to contain enough theory to fully understand the subject. This book will focus directly on the area of digital design and will explain the necessary concepts to understand and solve contemporary and future problems in a manner directly applicable by practicing engineers and/or students. It is worth noting that everything in this book has been applied to a successful modern design.

1.1. THE BASICS

As the reader undoubtedly knows, the basic idea in digital design is to communicate information with signals representing 1s or 0s. Typically this involves sending and receiving a series of trapezoidal shaped voltage signals such as shown in Figure 1.1 in which a high voltage is a 1 and a low voltage is a 0. The conductive paths carrying the digital signals are known as *interconnects*. The interconnect includes the entire electrical pathway from the chip sending a signal to the chip receiving the signal. This includes the chip packages, connectors, sockets, as well as a myriad of additional structures. A group of interconnects is referred to as a *bus*. The region of voltage where a digital receiver distinguishes between a high and a low voltage is known as the *threshold region*. Within this region, the receiver will either switch high or switch low. On the silicon, the actual switching voltages vary with temperature, supply voltage, silicon process, and other variables. From the system designers point of view, there are usually high- and low-voltage thresholds, known as *Vih* and *Vil*, associated with the receiving silicon, above which and below which a high or low value can be guaranteed to be received under all conditions. Thus the designer must guarantee that the system can, under all conditions, deliver high voltages that do not, even briefly, fall below Vih, and low voltages that remain below Vil, in order to ensure the integrity of the data.

In order to maximize the speed of operation of a digital system, the timing uncertainty of a transition through the threshold region must be minimized. This means that the rise or fall time of the digital signal must be as fast as possible. Ideally, an infinitely fast edge rate would be used, although there are many practical problems that prevent this. Realistically, edge rates of a few

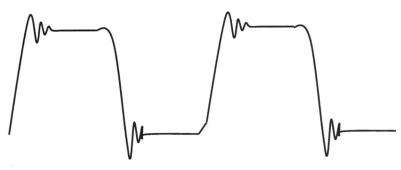

FIGURE 1.1   Digital waveform.

hundred picoseconds can be encountered. The reader can verify with Fourier analysis that the quicker the edge rate, the higher the frequencies that will be found in the spectrum of the signal. Herein lies a clue to the difficulty. Every conductor has a capacitance, inductance, and frequency-dependent resistance. At a high enough frequency, none of these things is negligible. Thus a wire is no longer a wire but a distributed parasitic element that will have delay and a transient impedance profile that can cause distortions and glitches to manifest themselves on the waveform propagating from the driving chip to the receiving chip. The wire is now an element that is coupled to everything around it, including power and ground structures and other traces. The signal is not contained entirely in the conductor itself but is a combination of all the local electric and magnetic fields around the conductor. The signals on one interconnect will affect and be affected by the signals on another. Furthermore, at high frequencies, complex interactions occur between the different parts of the same interconnect, such as the packages, connectors, vias, and bends. All these high-speed effects tend to produce strange, distorted waveforms that will indeed give the designer a completely different view of high-speed logic signals. The physical and electrical attributes of every structure in the vicinity of the interconnect has a vital role in the simple task of guaranteeing proper signaling transitions through Vih and Vil with the appropriate timings. These things also determine how much energy the system will radiate into space, which will lead to determining whether the system complies with governmental emission requirements. We will see in later chapters how to account for all these things.

When a conductor must be considered as a distributed series of inductors and capacitors, it is known as a *transmission line*. In general, this must be done when the physical size of the circuit under consideration approaches the wavelength of the highest frequency of interest in the signal. In the digital realm, since edge rate pretty much determines the maximum frequency content, one can compare rise and fall times to the size of the circuit instead, as shown in Figure 1.2. On a typical circuit board, a signal travels about half the speed of light (exact formulas will be in later chapters). Thus a 500 ps edge rate occupies about 3 in. in length on a circuit trace. Generally, any circuit length at least 1/10th of the edge rate must be considered as a transmission line.

One of the most difficult aspects of high-speed design is the fact that there are a large number codependent variables that affect the outcome of a digital design. Some of the variables are controllable and some force the designer to live with the random variation. One of the difficulties in high-speed design is how to handle the many variables, whether they are controllable or uncontrollable. Often simplifications can be made by neglecting or assuming values for variables, but this can lead to unknown failures down the road that will be impossible to "root cause" after the fact. As timing becomes more constrained, the simplifications of the past are rapidly dwindling in utility to the modern designer. This book will also show how to incorporate a large num-

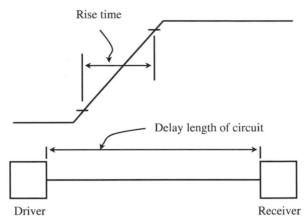

FIGURE 1.2   Rise time and circuit length.

ber of variables that would otherwise make the problem intractable. Without a methodology for handling the large amount of variables, a design ultimately resorts to guesswork no matter how much the designer physically understands the system. The final step of handling all the variables is often the most difficult part and the one most readily ignored by a designer. A designer crippled by an inability to handle large amounts of variables will ultimately resort to proving a few "point solutions" instead and hope that they plausibly represent all known conditions. While sometimes such methods are unavoidable, this can be a dangerous guessing game. Of course, a certain amount of guesswork is always present in a design, but the goal of the system designer should be to minimize uncertainty.

1.2. THE PAST AND THE FUTURE

Gordon Moore, co-founder of Intel Corporation, predicted that the perfor- mance of computers will double every 18 months. History confirmed this insightful prediction. Remarkably, computer performance has doubled approx- imately every 1.5 years, along with substantial decreases in their price. One measure of relative processor performance is internal clock rates. Figure 1.3 shows several processors through history and their associated internal clock rates. By the time this is in print, even the fastest processors on this chart will likely be considered unimpressive. The point is that computer speeds are increasing exponentially. As core frequency increases, faster data rates will be demanded from the buses that feed information to the processor, as shown in Figure 1.4, leading to an interconnect timing budget that is decreasing ex- ponentially. Decreased timing budgets mean that it is evermore important to properly account for any phenomenon that may increase the timing uncertainty of the digital waveform as it arrives at the receiver. This is the root cause of

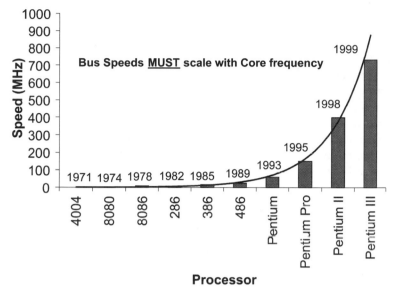

FIGURE 1.3 Moore's law in action.

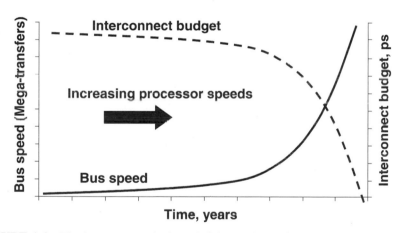

FIGURE 1.4 The interconnect budget shrinks as the performance and frequency of the system increases.

two inescapable obstacles that will continue to make digital system design difficult. The first obstacle is simply that the sheer amount of variables that must be accounted for in a digital design is increasing. As frequencies increase, new effects, which may have been negligible at slower speeds, start to become significant. Generally speaking, the complexity of a design increases exponentially with increasing variable count. The second obstacle is that the new effects, which could be ignored in designs of the past, must be modeled to a very high precision. Often these new models are required to be three-dimensional in nature, or require specialized analog techniques that fall outside the realms of the digital designer's discipline. The obstacles are perhaps more profound on the subsystems surrounding the processor since they evolve at a much slower rate, but still must support the increasing demands of the processor.

All of this leads to the present situation: There are new problems to solve. Engineers who can solve these problems will define the future. This book will equip the reader with the necessary practical understanding to contend with modern high-speed digital design and with enough theory to see beyond this book and solve problems that the authors have not yet encountered. Read on.

CHAPTER 2

Ideal Transmission Line Fundamentals

In today's high-speed digital systems, it is necessary to treat the printed circuit board (PCB) or multichip module (MCM) traces as transmission lines. It is no longer possible to model interconnects as lumped capacitors or simple delay lines, as could be done on slower designs. This is because the timing issues associated with the transmission lines are becoming a significant percentage of the total timing margin. Great attention must be given to the construction of the PCB so that the electrical characteristics of the transmission lines are controlled and predictable. In this chapter we introduce the basic transmission line structures typically used in digital systems and present basic transmission line theory for the ideal case. The material presented in this chapter provides the necessary knowledge base needed to comprehend all subsequent chapters.

2.1. TRANSMISSION LINE STRUCTURES ON A PCB OR MCM

Transmission line structures seen on a typical PCB or MCM consist of conductive traces buried in or attached to a dielectric or insulating material with one or more reference planes. The metal in a typical PCB is usually copper and the dielectric is FR4, which is a type of fiberglass. The two most common types of transmission lines used in digital designs are microstrips and striplines. A *microstrip* is typically routed on an outside layer of the PCB and has only one reference plane. There are two types of microstrips, buried and nonburied. A *buried* (sometimes called *embedded*) *microstrip* is simply a transmission line that is embedded into the dielectric but still has only one reference plane. A *stripline* is routed on an inside layer and has two reference planes. Figure 2.1 represents a PCB with traces routed between the various components on both internal (stripline) and external (microstrip) layers. The accompanying cross section is taken at the given mark so that the position of transmission lines relative to the ground/power planes can be seen. In this book, transmission lines are often represented in the form of a cross section. This is very useful for calculating and visualizing the various transmission line parameters described later.

Multiple-layer PCBs such as the one depicted in Figure 2.1 can provide a variety of stripline and microstrip structures. Control of the conductor and

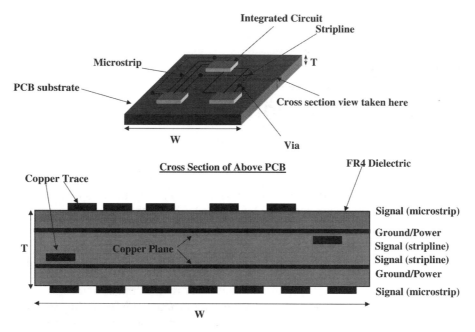

FIGURE 2.1   Example transmission lines in a typical design built on a PCB.

dielectric layers (which is referred to as the *stackup*) is required to make the electrical characteristics of the transmission line predictable. In high-speed systems, control of the electrical characteristics of the transmission lines is crucial. These basic electrical characteristics, defined in this chapter, will be referred to as *transmission line parameters*.

2.2. WAVE PROPAGATION

At high frequencies, when the edge rate (rise and fall times) of the digital signal is small compared to the propagation delay of an electrical signal traveling down the PCB trace, the signal will be greatly affected by transmission line effects. The electrical signal will travel down the transmission line in the way that water travels through a long square pipe. This is known as *electrical wave propagation*. Just as the waterfront will travel as a wave down the pipe, an electrical signal will travel as a wave down a transmission line. Additionally, just as the water will travel the length of the pipe in a finite amount of time, the electrical signal will travel the length of the transmission line in a finite amount of time. To take this simple analogy one step further, the voltage on a transmission line can be compared to the height of the water in the pipe, and the flow of the water can be compared to the current. Figure 2.2 depicts a common way of representing a transmission line. The top line is the signal

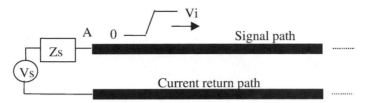

FIGURE 2.2 Typical method of portraying a digital signal propagating on a transmission line.

path and the bottom line is the current return path. The voltage V_i is the initial voltage launched onto the line at node A, and V_s and Z_s form a Thévenin equivalent representation of the output buffer, usually referred to as the *source* or the *driver*.

2.3. TRANSMISSION LINE PARAMETERS

To analyze the effects that transmission lines have on high speed digital systems, the electrical characteristics of the line must be defined. The basic electrical characteristics that define a transmission line are its characteristic impedance and its propagation velocity. The *characteristic impedance* is similar to the width of the water pipe used in the analogy above, and the *propagation velocity* is simply analogous to speed at which the water flows through the pipe. To define and derive these terms, it is necessary to examine the fundamental properties of a transmission line. As a signal travels down the transmission line depicted in Figure 2.2, there will be a voltage differential between the signal path and the current return path (generically referred to as a *ground return path* or an *ac ground* even when the reference plane is a power plane). When the signal reaches an arbitrary point z on the transmission line, the signal path conductor will be at a potential of V_i volts and the ground return conductor will be at a potential of 0 V. This voltage difference establishes an electric field between the signal and the ground return conductors. Furthermore, Ampère's law states that the line integral of the magnetic field taken about any given closed path must be equal to the current enclosed by that path. In simpler terms, this means that if a current is flowing through a conductor, it results in a magnetic field around that conductor. We have therefore established that if an output buffer injects a signal of voltage V_i and current I_i onto a transmission line, it will induce an electric and a magnetic field, respectively. However, it should be clear that the voltage V_i and current I_i, at any arbitrary point on the line z will be zero until the time z/v, where v is the velocity of the signal traveling down the transmission line and z is the distance from the source. Note that this analysis implies that the signal is not simply traveling on the signal conductor of the transmission line; rather, it is traveling between the sig-

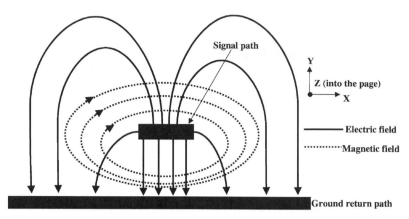

FIGURE 2.3 Cross section of a microstrip depicting the electric and magnetic fields assuming that an electrical signal is propagating down the line into the page.

nal conductor and reference plane in the form of an electric and a magnetic field.

Now that the basic electromagnetic properties of a transmission line have been established, it is possible to construct a simple circuit model for a section of the line. Figure 2.3 represents a cross section of a microstrip transmission line and the electric and magnetic field patterns associated with a current flowing though the line. If it is assumed that there are no components of the electric or magnetic fields propagating in the z-direction (into the page), the electric and magnetic fields will be orthogonal. This is known as *transverse electromagnetic mode* (TEM). Transmission lines will propagate in TEM mode under normal circumstances and it is an adequate approximation even at relatively high frequencies. This allows us to examine the transmission line in differential sections (or slices) along the length of the line traveling in the z-direction (into the page). The two components shown in Figure 2.3 are the electric and magnetic fields for an infinitesimal or differential section (slice) of the transmission line of length dz. Since there is energy stored in both an electric and a magnetic field, let us include the circuit components associated with this energy storage in our circuit model. The magnetic field for a differential section of the transmission line can be represented by a series inductance $L\,dz$, where L is inductance per length. The electric field between the signal path and the ground path for a length of dz can be represented by a shunt capacitor $C\,dz$, where C is capacitance per length. An ideal model would consist of an infinite number of these small sections cascaded in series. This model adequately describes a section of a loss-free transmission line (i.e., a transmission line with no resistive losses).

However, since the metal used in PCB boards is not infinitely conductive and the dielectrics are not infinitely resistive, loss mechanisms must be added to the model in the form of a series resistor, $R\,dz$, and a shunt resistor to

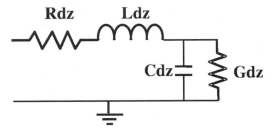

FIGURE 2.4 Equivalent circuit model of a differential section of a transmission line of length dz (*RLCG* model).

ground referred to as a conductance, Gdz, with units of siemens (1/ohm). Figure 2.4 depicts the equivalent circuit model for a differential section of a transmission line. The series resistor, Rdz, represents the losses due to the finite conductivity of the conductor; the shunt resistor, Gdz, represents the losses due to the finite resistance of the dielectric separating the conductor and the ground plane, the series inductor, Ldz, represents the magnetic field; and the capacitor, Cdz, represents the electric field between the conductor and the ground plane. In the remainder of this book, one of these sections will be known as an *RLCG element*.

2.3.1. Characteristic Impedance

The characteristic impedance Z_o of the transmission line is defined by the ratio of the voltage and current waves at any point of the line; thus, $V/I = Z_o$. Figure 2.5 depicts two representations of a transmission line. Figure 2.5*a*

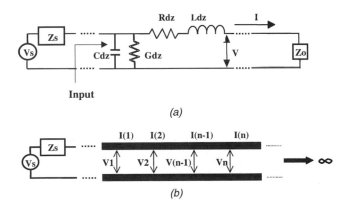

FIGURE 2.5 Method of deriving a transmission lines characteristic impedance: (*a*) differential section; (*b*) infinitely long transmission line.

represents a differential section of a transmission line of length dz modeled with an *RLCG* element as described above and terminated in an impedance of Z_o. The characteristic impedance of the *RLCG* element is defined as the ratio of the voltage V and current I, as depicted in Figure 2.5a. Assuming that the load Z_o is exactly equal to the characteristic impedance of the *RLCG* element, Figure 2.5a can be represented by Figure 2.5b, which is an infinitely long transmission line. The termination, Z_o, in Figure 2.5a simply represents the infinite number of additional *RLCG* segments of impedance Z_o that comprise the complete transmission line model. Since the voltage/current ratio in the terminating device, Z_o, will be the same as that in the *RLCG* segment, then from the perspective of the voltage source, Figure 2.5a and b will be indistinguishable. With this simplification, the characteristic impedance can be derived for an infinitely long transmission line.

To derive the characteristic impedance of the line, Figure 2.5a should be examined. Solving the equivalent circuit of Figure 2.5a for the input impedance with the assumption that the characteristic impedance of the line is equal to the terminating impedance, Z_o, yields equation (2.1). For simplicity, the differential length dz is replaced with a short length of Δz. The derivation is as follows: Let

$$j\omega L(\Delta z) + R(\Delta z) = Z\Delta z \qquad \text{(series impedance for length of line } \Delta z\text{)}$$

$$j\omega C(\Delta z) + G(\Delta z) = Y\Delta z \qquad \text{(parallel admittance for length of line } \Delta z\text{)}$$

Then

$$Z(\text{input}) = Z_o = \frac{(Z_o + Z\,\Delta z)(1/Y\,\Delta z)}{Z_o + Z\Delta z + 1/Y\,\Delta z} \qquad \text{(assuming the load is equal to the characteristic impedance)}$$

$$Z_o\left(Z\,\Delta z + Z_o + \frac{1}{Y\,\Delta z}\right) = (Z_o + Z\,\Delta z)\frac{1}{Y\,\Delta z}$$

$$\Rightarrow Z_o Z\,\Delta z + Z_o^2 + \frac{Z_o}{Y\,\Delta z} = \frac{Z_o}{Y\,\Delta z} + \frac{Z\,\Delta z}{Y\,\Delta z}$$

$$\Rightarrow Z_o(Z\,\Delta z + Z_o) = \frac{Z}{Y}$$

$$\Rightarrow Z_o Y(Z\,\Delta z + Z_o) = Z$$

$$\Rightarrow \lim_{\Delta z \to 0}[Z] = Z_o^2 Y$$

Therefore,

$$Z_o = \sqrt{\frac{Z}{Y}} = \sqrt{\frac{R + j\omega L}{G + j\omega C}} \tag{2.1}$$

where R is in ohms per unit length, L is in henries per unit length, G is in siemens per unit length, C is in farads per unit length, and ω is in radians per second. It is usually adequate to approximate the characteristic impedance as $Z_o = \sqrt{L/C}$, since R and G both tend to be significantly smaller than the other terms. Only at very high frequencies, or with very lossy lines, do the R and G components of the impedance become significant. (Lossy transmission lines are covered in Chapter 4). Lossy lines will also yield complex characteristic impedances (i.e., having imaginary components). For the purposes of digital design, however, only the magnitude of the characteristic impedance is important.

For maximum accuracy, it is necessary to use one of the many commercially available two-dimensional electromagnetic field solvers to calculate the impedance of the PCB traces for design purposes. The solvers will typically provide the impedance, propagation velocity, and L and C elements per unit length. This is adequate since R and G usually have a minimal effect on the impedance. In the absence of a field solver, the formulas presented in Figure 2.6 will provide good approximations to the impedance values of typical transmission lines as a function of the trace geometry and the dielectric constant (ε_r). More accurate formulas for characteristic impedance are presented in Appendix A.

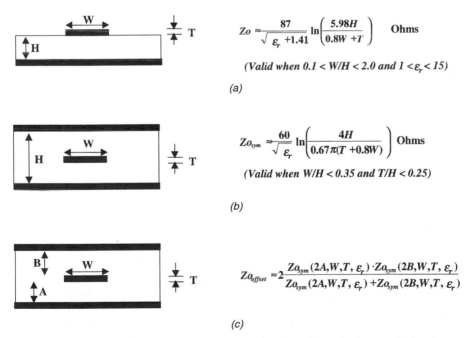

$$Zo \approx \frac{87}{\sqrt{\varepsilon_r + 1.41}} \ln\left(\frac{5.98H}{0.8W + T}\right) \quad \textbf{Ohms}$$

(Valid when 0.1 < W/H < 2.0 and 1 < ε_r < 15)

(a)

$$Zo_{sym} \approx \frac{60}{\sqrt{\varepsilon_r}} \ln\left(\frac{4H}{0.67\pi(T + 0.8W)}\right) \quad \textbf{Ohms}$$

(Valid when W/H < 0.35 and T/H < 0.25)

(b)

$$Zo_{offset} \approx 2 \frac{Zo_{sym}(2A,W,T,\varepsilon_r) \cdot Zo_{sym}(2B,W,T,\varepsilon_r)}{Zo_{sym}(2A,W,T,\varepsilon_r) + Zo_{sym}(2B,W,T,\varepsilon_r)}$$

(c)

FIGURE 2.6 Characteristic impedance approximations for typical transmission lines: (*a*) microstrip line; (*b*) symmetrical stripline; (*c*) offset stripline.

2.3.2. Propagation Velocity, Time, and Distance

Electrical signals on a transmission line will propagate at a speed that depends on the surrounding medium. Propagation delay is usually measured in terms of seconds per meter and is the inverse of the propagation velocity. The propagation delay of a transmission line will increase in proportion to the square root of the surrounding dielectric constant. The time delay of a transmission line is simply the amount of time it takes for a signal to propagate the entire length of the line. The following equations show the relationships between the dielectric constant, the propagation velocity, the propagation delay, and the time delay:

$$v = \frac{c}{\sqrt{\varepsilon_r}} \tag{2.2}$$

$$PD = \frac{1}{v} = \frac{\sqrt{\varepsilon_r}}{c} \tag{2.3}$$

$$TD = \frac{x\sqrt{\varepsilon_r}}{c} \tag{2.4}$$

where

v = propagation velocity, in meters/second

c = speed of light in a vacuum (3×10^8 m/s)

ε_r = dielectric constant

PD = propagation delay, in seconds per meter

TD = time delay for a signal to propagate down a transmission line of length x

x = length of the transmission line, in meters

The time delay can also be determined from the equivalent circuit model of the transmission line:

$$TD = \sqrt{LC} \tag{2.5}$$

where L is the total series inductance for the length of the line and C is the total shunt capacitance for the length of the line.

It should be noted that equations (2.2) through (2.4) assume that no magnetic materials are present, such that $\mu_r = 1$, and thus effects due to magnetic materials can be left out of the formulas.

The delay of a transmission line depends on the dielectric constant of the dielectric material, the line length, and the geometry of the transmission line cross section. The cross-sectional geometry determines whether the electric field will stay completely contained within the board or fringe out into the air. Since a typical PCB board is made out of FR4, which has a dielectric constant of approximately 4.2, and air has a dielectric constant of 1.0, the resulting "effective" dielectric constant will be a weighted average between the two. The amount of the electric field that is in the FR4 and the amount that is in the air determine the effective value. When the electric field is completely contained within the board, as in the case of a stripline, the effective dielectric constant will be larger and the signals will propagate more slowly than will externally routed traces. When signals are routed on the external layers of the board as in the case of a microstrip line, the electric field fringes through the dielectric material and the air, which lowers the effective dielectric constant; thus the signals will propagate more quickly than those on an internal layer.

The effective dielectric constant for a microstrip is calculated as follows [Collins, 1992]:

$$\varepsilon_e = \frac{\varepsilon_r + 1}{2} + \frac{\varepsilon_r - 1}{2}\left(1 + \frac{12H}{W}\right)^{-1/2} + F - 0.217(\varepsilon_r - 1)\frac{T}{\sqrt{WH}} \quad (2.6)$$

$$F = \begin{cases} 0.02(\varepsilon_r - 1)\left(1 - \dfrac{W}{H}\right)^2 & \text{for } \dfrac{W}{H} < 1 \\ \\ 0 & \text{for } \dfrac{W}{H} > 1 \end{cases} \quad (2.7)$$

where ε_r is the dielectric constant of the board material, H the height of the conductor above the ground plane, W the conductor width, and T the conductor thickness.

2.3.3. Equivalent Circuit Models for SPICE Simulation

In Section 2.3 we introduced the equivalent distributed circuit model of a transmission line, which consisted of an infinite number of *RLCG* segments cascaded together. Since it is not practical to model a transmission line with an infinite number of elements, a sufficient number can be determined based on the minimum rise or fall time used in the simulation. When simulating a digital system, it is usually sufficient to choose the values so that the time delay (TD = $\sqrt{LC}$) of the shortest *RLCG* segment is no larger than one-tenth of the minimum system rise or fall time. The rise or fall time is defined as the amount of time it takes a signal to transition between its minimum and maximum magnitude. Rise times are typically measured between the 10 and 90% values of the maximum swing. For example, if a signal transitioned from 0 V to 1 V, its rise time would be measured between the times when the voltage reaches 0.1 and 0.9 V.

RULE OF THUMB: Choosing a Sufficient Number of *RLCG* Segments

When using a distributed *RLCG* model for modeling transmission lines, the number of *RLCG* segments should be determined as follows:

$$\text{segments} \geq 10 \left(\frac{x}{T_r v} \right)$$

where x is the length of the line, v the propagation velocity of the transmission line, and T_r the rise (or fall) time. Each parasitic in the model should be scaled by the number of segments. For example, if the parasitics are known per unit meter, the maximum values used for a single segment must be

$$C_{\text{segment}} = \frac{(x)(C/\text{meter})}{\text{segments}}$$

$$L_{\text{segment}} = \frac{(x)L/\text{meter}}{\text{segments}}$$

$$R_{\text{segment}} = \frac{(x)R/\text{meter}}{\text{segments}}$$

$$G_{\text{segment}} = \frac{(x)G/\text{meter}}{\text{segments}}$$

$$\text{TD}_{\text{segment}} = \sqrt{L_{\text{segment}}C_{\text{segment}}} \leq \frac{T_r}{10}$$

Example 2.1: Creating a Transmission Line Model. Create an equivalent circuit model of a loss-free 50-Ω transmission line 5 in. long for the cross section shown in Figure 2.7a. Assume that the driver has a minimum rise time of 2.5 ns. Assume a dielectric constant of 4.5.

SOLUTION: Initially, the inductance and capacitance of the transmission line must be calculated. Since no field solver is available, the equations presented above will be used.

$$Z_o \approx \frac{60}{\sqrt{\varepsilon_r}} \ln \frac{4H}{0.67\pi(T + 0.8W)} = \frac{60}{\sqrt{\varepsilon_r}} \ln \frac{4(14.7)}{0.67\pi[0.7 + 0.8(5)]} = 50 \ \Omega$$

$$\text{TD} = \frac{x\sqrt{\varepsilon_r}}{c} = 5 \text{ in.}(0.0254 \text{ m/in.})\frac{\sqrt{4.5}}{3 \times 10^8 \text{ m/s}} = 898 \text{ ps}$$

$$v = \frac{c}{\sqrt{\varepsilon_r}} = \frac{3 \times 10^8 \text{ m/s}}{\sqrt{4.5}} = 1.41 \times 10^8 \text{ m/s}$$

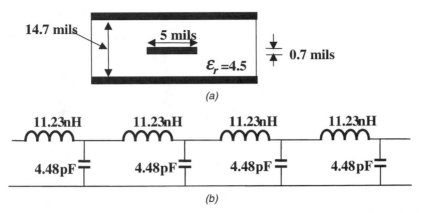

(a)

(b)

FIGURE 2.7 Creating a transmission line model: (a) cross section; (b) equivalent circuit.

If the transmission line is a microstrip, the same procedure is used to calculate the velocity, but with the effective dielectric constant as calculated in equation (2.6).

Since $Z_o = \sqrt{L/C}$ and TD $= \sqrt{LC}$, we have two equations and two unknowns. Solve for L and C.

$$L_{\text{total}} = (\text{TD})(Z_o) = (898 \times 10^{-12})(50\ \Omega) = 44.9\ \text{nH}$$

$$C_{\text{total}} = \frac{\text{TD}}{Z_o} = \frac{898 \times 10^{-12}\ \text{s}}{50\ \Omega} = 17.9\ \text{pF}$$

The L and C values above are the total inductance and capacitance for the 5-in. line.

$$\text{segments} \geq 10\left(\frac{X}{T_r v}\right) = 10\left[\frac{5\ \text{in.}(0.0254\ \text{m/in.})}{2.5\ \text{ns}(1.41 \times 10^8\ \text{m/s})}\right] = 3.6$$

Because 3.6 is not a round number, we will use four segments in the model.

$$C_{\text{segment}} = \frac{C_{\text{total}}}{\text{segments}} = \frac{17.9\ \text{pF}}{4} = 4.48\ \text{pF}$$

$$L_{\text{segment}} = \frac{L_{\text{total}}}{\text{segments}} = \frac{44.9\ \text{nH}}{4} = 11.23\ \text{nH}$$

The final loss-free transmission line equivalent circuit is shown in Figure 2.7b.

Double check to ensure that the rule of thumb is satisfied.

$$\text{TD}_{\text{segment}} = \sqrt{L_{\text{segment}}C_{\text{segment}}} = \sqrt{(4.48 \text{ pF})(11.23 \text{ nH})} = 0.224 \text{ ns} \leq \frac{T_r}{10}$$

2.4. LAUNCHING INITIAL WAVE AND TRANSMISSION LINE REFLECTIONS

The characteristics of the driving circuitry and the transmission line greatly affect the integrity of a signal being transmitted from one device to another. Subsequently, it is very important to understand how the signal is launched onto a transmission line and how it will look at the receiver. Although many parameters will affect the integrity of the signal at the receiver, in this section we describe the most basic behavior.

2.4.1. Initial Wave

When a driver launches a signal onto a transmission line, the magnitude of the signal depends on the voltage and source resistance of the buffer and the impedance of the transmission line. The initial voltage seen at the driver will be governed by the voltage divider of the source resistance and the line impedance. Figure 2.8 depicts an initial wave being launched onto a long transmission line. The initial voltage V_i will propagate down the transmission line until it reaches the end. The magnitude of V_i is determined by the voltage divider between the source and the line impedance:

$$V_i = V_s \frac{Z_o}{Z_o + Z_s} \tag{2.8}$$

If the end of the transmission line is terminated with an impedance that exactly matches the characteristic impedance of the line, the signal with amplitude V_i will be terminated to ground and the voltage V_i will remain on the line until the signal source switches again. In this case the voltage V_i is the dc steady-state value. Otherwise, if the end of the transmission line exhibits some impedance other than the characteristic impedance of the line, a portion of the signal will be terminated to ground and the remainder of the sig-

FIGURE 2.8   Launching a wave onto a long transmission line.

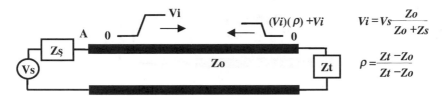

FIGURE 2.9   Incident signal being reflected from an unmatched load.

nal will be reflected back down the transmission line toward the source. The amount of signal reflected back is determined by the *reflection coefficient*, defined as the ratio of the reflected voltage to the incident voltage seen at a given junction. In this context, a *junction* is defined as an impedance discontinuity on a transmission line. The impedance discontinuity could be a section of transmission line with different characteristic impedance, a terminating resistor, or the input impedance to a buffer on a chip. The reflection coefficient is calculated as

$$\rho = \frac{V_{\text{reflected}}}{V_{\text{incident}}} = \frac{Z_t - Z_o}{Z_t + Z_o} \tag{2.9}$$

impedance of the line, and Z_t the impedance of the discontinuity. The equation assumes that the signal is traveling on a transmission line with characteristic impedance Z_o and encounters an impedance discontinuity of Z_t. Note that if $Z_o = Z_t$, the reflection coefficient is zero, meaning that there is no reflection. The case where $Z_o = Z_t$ is known as *matched termination.*

As depicted in Figure 2.9, when the incident wave hits the termination Z_t, a portion of the signal, $V_i\rho$, is reflected back toward the source and is added to the incident wave to produce a total magnitude on the line of $V_i\rho + V_i$. The reflected component will then travel back to the source and possibly generate another reflection off the source. This reflection and counterreflection continues until the line has reached a stable condition.

Figure 2.10 depicts special cases of the reflection coefficient. When the line is terminated in a value that is exactly equal to its characteristic impedance, there is no discontinuity, and the signal is terminated to ground with no reflections. With open and shorted loads, the reflection is 100%, however, the reflected signal is positive and negative, respectively.

2.4.2. Multiple Reflections

As described above, when a signal is reflected from an impedance discontinuity at the end of the line, a portion of the signal will be reflected back toward the source. When the reflected signal reaches the source, another reflection

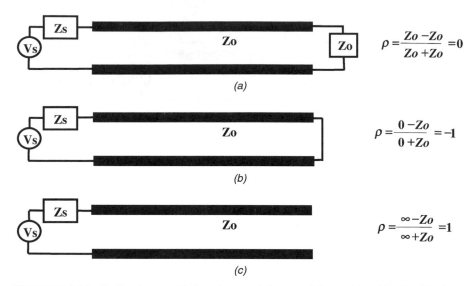

FIGURE 2.10   Reflection coefficient for special cases: (*a*) terminated in Z_o; (*b*) short circuit; (*c*) open circuit.

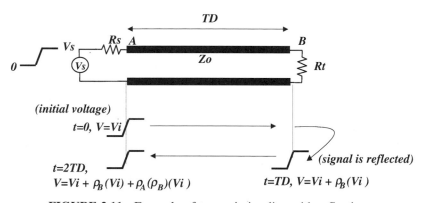

FIGURE 2.11   Example of transmission line with reflections.

will be generated if the source impedance does not equal that of the transmission line. Subsequently, if an impedance discontinuity exists on both sides of the transmission line, the signal will bounce back and forth between the driver and receiver. The signal reflections will eventually reach steady state at the dc solution.

For example, consider Figure 2.11, which shows one example for a time interval of a few TD (where TD is the time delay of the transmission line from source to load). When the source transitions to V_s, the initial voltage on the line, V_i, is determined by the voltage divider $V_i = V_s Z_o/(Z_o + R_s)$. At time t = TD, the incident voltage V_i arrives at the load R_t. At this time a re-flected component is generated with a magnitude of $\rho_B V_i$, which is added to the incident voltage V_i, creating a total voltage at the load of $V_i + \rho_B V_i$ (ρ_B is the reflection coefficient looking into the load). The reflected portion of the wave ($\rho_B V_i$) then travels back to the source and at time t = 2TD generates a reflection off the source determined by $\rho_A \rho_B V_i$ (ρ_A is the reflection coef-ficient looking into the source). At this time the voltage seen at the source will be the previous voltage (V_i) plus the incident transient voltage from the reflection ($\rho_B V_i$) plus the reflected wave ($\rho_A \rho_B V_i$). This reflecting and counter-reflecting will continue until the line voltage has approached the steady-state dc value. As the reader can see, the reflections could take a long time to settle out if the termination is not matched and can have some significant timing impacts.

It is apparent that hand calculation of multiple reflections can be rather tedious. An easier way to predict the effect of reflections on a signal is to use a lattice diagram.

Lattice Diagrams and Over- and Underdriven Transmission Lines. A *lattice diagram* (sometimes called a *bounce diagram*) is a technique used to solve the multiple reflections on a transmission line with linear loads. Figure 2.12 shows a sample lattice diagram. The left- and right-hand vertical lines represent the source and load ends of the transmission line. The diagonal lines contained between the vertical lines represent the signal bouncing back and forth between the source and the load. The diagram progressing from top to bottom represents increasing time. Notice that the time increment is equal to the time delay of the transmission line. Also note that the vertical bars are labeled with reflection coefficients at the top of the diagram. These reflection coefficients represent the reflection between the transmission line and the load (looking into the load from the line) and the reflection coefficient looking into the source. The lowercase letters represent the magnitude of the reflected signal traveling on the line, the uppercase letters represent the voltages seen at the source, and the primed uppercase letters represent the voltage seen at the load end of the line. For example, referring to Figure 2.12, the near end of the line will be held at a voltage of A volts for a duration of $2N$ picoseconds, where N is the time delay (TD) of the transmission line. The voltage A is simply the initial voltage $V_{initial}$, which will remain constant until the reflection from the load reaches the source. The voltage A' is simply the voltage a plus the reflected voltage b. The voltage B is the sum of the incident voltage a, the signal reflected from the load b, the signal reflected off the source c, and so on. The reflections on the line eventually reach the steady-state voltage of the source, V_s, if the line is open. However, if the line is terminated with a resistor, R_t, the steady-state voltage is computed as

$$V_s \frac{R_t}{R_t + R_s} \qquad (2.10)$$

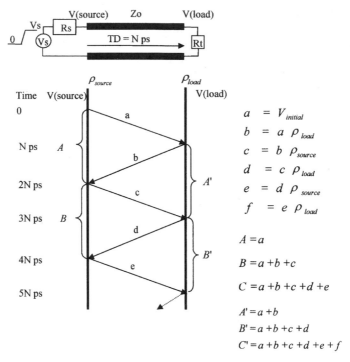

FIGURE 2.12 Lattice diagram used to calculate multiple reflections on a transmission line.

Example 2.2: Multiple Reflections for an Underdriven Transmission Line. As described above, when the driver launches a signal onto the transmission line, the initial voltage present on the transmission line will be governed by the voltage divider between the driver impedance Z_s and the line impedance Z_0. As shown in Figure 2.13, this value is 0.8 V. The initial signal, 0.8 V, will travel down the line until it reaches the load. In this particular case, the load is open and thus has a reflection coefficient of 1. Subsequently, the entire signal is reflected back toward the source and is added to the incident signal of 0.8 V. So at time = TD, 250 ps in this example, the signal seen at the load is $0.8 + 0.8$, or 1.6 V. The 0.8-V reflected signal will then propagate down the line toward the source. When the signal reaches the source, part of the signal will be reflected back toward the load. The magnitude of the reflected signal depends on the reflection coefficient between the line impedance Z_0 and the source impedance Z_s. In this example the value reflected toward the load is (0.8 V)(0.2), which is 0.16 V. The reflected signal will be added to the signal already present on the line, which will give a total magnitude of 1.76 V, with the reflected portion of 0.16 V traveling to the load. This process is repeated until the voltage reaches a steady-state value of 2 V.

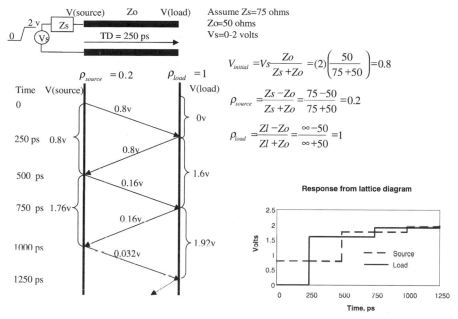

FIGURE 2.13 Example 2.2: Lattice diagram used to calculate multiple reflections for an underdriven transmission line.

The response of the lattice diagram is shown in the lower corner of Figure 2.13. A computer simulation of the response is shown in Figure 2.14 for comparison. Notice how the reflections give the waveform a "stair-step" appearance at the receiver, even though the unloaded output of the voltage source is a square wave. This effect occurs when the source impedance Z_s is larger than the line impedance Z_o and is referred to as an *underdriven transmission line*.

Example 2.3: Multiple Reflections for an Overdriven Transmission Line. When the line impedance is greater than the source impedance, the reflection coefficient looking into the source will be negative, which will produce a "ringing" effect. This is known as an *overdriven transmission line*. The lattice diagram for an overdriven transmission line is shown in Figure 2.15. Figure 2.16 is a SPICE simulation showing the response of the system depicted in Figure 2.15.

Next, consider the transmission line structure depicted in Figure 2.17. The structure consists of two segments of transmission line cascaded in series. The first section is of length X and has a characteristic impedance of Z_{o1} ohms. The second section is also of length X and has an impedance of Z_{o2} ohms. Finally, the structure is terminated with a value of R_t. When the signal encounters the Z_{o1}/Z_{o2} impedance junction, part of the signal will be reflected, as governed

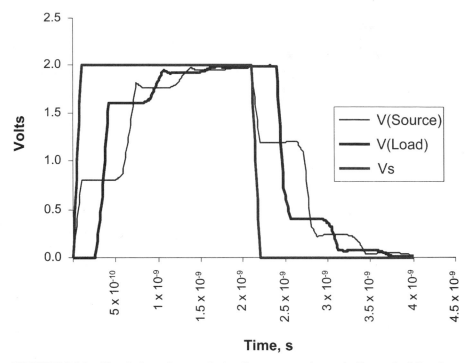

FIGURE 2.14 Simulation of transmission line system shown in Example 2.2, where the line impedance is less than the source impedance (underdriven transmission line).

by the reflection coefficient, and part of the signal will be transmitted, as governed by the transmission coefficient:

$$T = 1 + \rho \tag{2.11}$$

Figure 2.17 also depicts how a lattice diagram can be used to solve for multiple reflections on a transmission line system with more than one characteristic impedance. Note that the transmission lines in this example are of equal length, which simplifies the problem because the reflections on each section will be in phase. For example, refer to Figure 2.17 and note that the reflection, e, adds directly to the reflection, f. When the two transmission lines are of different lengths, the reflections from one section will not be in phase with the reflections from the other section, which complicates the diagram drastically. Once the system complexity progresses beyond the point depicted in Figure 2.17, it is preferable to use a simulator such as SPICE to solve the system.

Bergeron Diagrams and Reflections from Nonlinear Loads. The *Bergeron diagram* is another technique used for solving multiple reflections on a

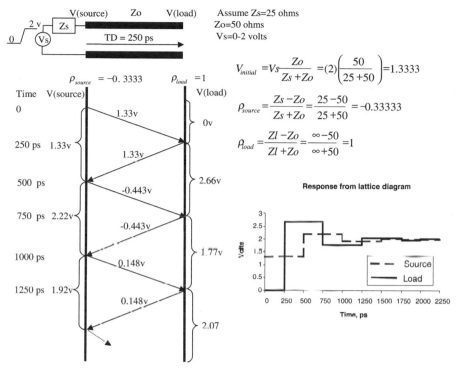

FIGURE 2.15 Example 2.3: Lattice diagram used to calculate multiple reflections for an overdriven transmission line.

transmission line. A Bergeron diagram is used in place of a lattice diagram when nonlinear loads and sources exist in the system. A good example of when a Bergeron diagram is necessary is when a transmission line is terminated with a clamping diode to prevent excess signal overshoot or damage due to electrostatic discharge. Furthermore, output buffers rarely exhibit perfectly linear I–V characteristics; thus a Bergeron diagram will give a more accurate representation of the reflections if the I–V characteristics of the buffers are well known.

Refer to Figure 2.18. To construct a Bergeron diagram, plot the I–V characteristics of the load and the source. The source I–V curve will have a negative slope of $1/R_s$ since current is leaving the node and the X intercept will be at V_s. Then, beginning with the transmission line's initial condition (i.e., $V = 0$, $I = 0$), construct a line with a slope of $1/Z_o$. The point where this line intersects the source I–V curve gives the initial voltage and current on the line at the source at time = 0. You may recognize this as a load diagram. From the intersection with the source line, construct a line with a slope of $-1/Z_o$ and extend the line to the load curve. This vector represents the signal traveling down the line toward the load. The intersection with the load line will define

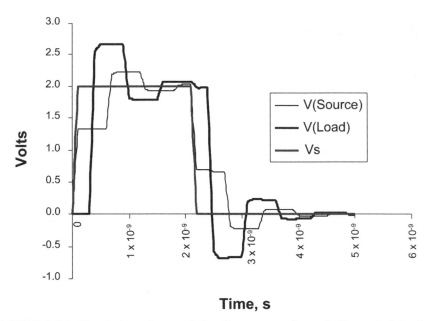

FIGURE 2.16 Simulation of transmission line system shown in Example 2.3 where the line impedance is greater than the source impedance (over-driven transmission line).

the voltage and current at the load at time = TD, where TD is the time delay of the line. Repeat this procedure using alternating slopes of $1/Z_o$ and $-1/Z_o$ until the transmission line vectors reach the point where the load and source lines intersect. The intersections of the transmission line vectors and the load and source I–V curves give the voltage and current values at steady state. Figure 2.19 is an example that calculates the response of a similar system where $V_s = 3$ V, TD = 500 ps, $Z_o = 50$ Ω, $R_s = 25$ Ω, and the diode behaves with the equation shown.

POINT TO REMEMBER

Use a Bergeron diagram to calculate the reflections on a transmission line when either the source or load exhibits significant nonlinear I–V characteristics.

2.4.3. Effect of Rise Time on Reflections

The rise time will begin to have a significant effect on the wave shape when it becomes less than twice the delay (TD) of the transmission line. Figures 2.20

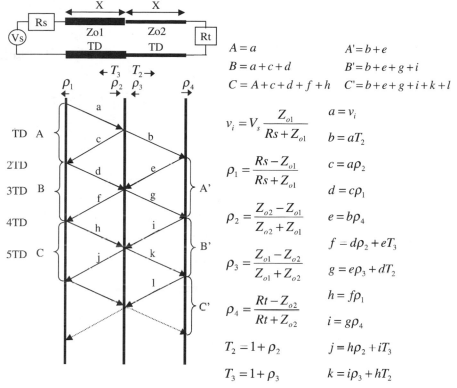

FIGURE 2.17   Lattice diagram of transmission line system with multiple line impedances.

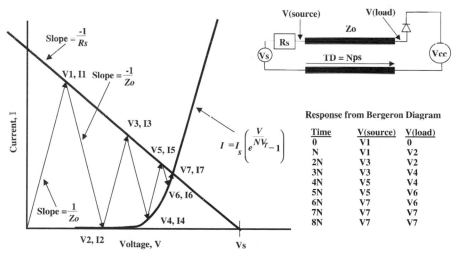

FIGURE 2.18   Bergeron diagram used to calculate multiple reflection with a nonlinear load.

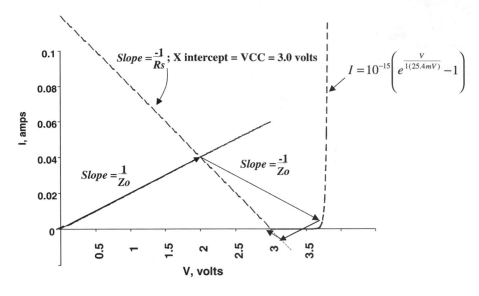

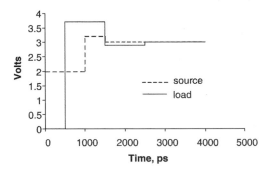

FIGURE 2.19 Bergeron diagram used to calculate the reflection on a transmission line with a diode termination.

and 2.21 show the effect that edge rate has on underdriven and overdriven transmission lines. Notice how significantly the wave shape changes as the rise time exceeds twice the delay of the line. When the edge rate exceeds twice the line delay, the reflection from the source arrives before the transition from one state to another is complete (i.e., high-to-low or low-to-high transition).

2.4.4. Reflections from Reactive Loads

In real systems there are rarely cases where the loads are purely resistive. The input to a CMOS gate, for example, tends to be capacitive. Additionally, bond

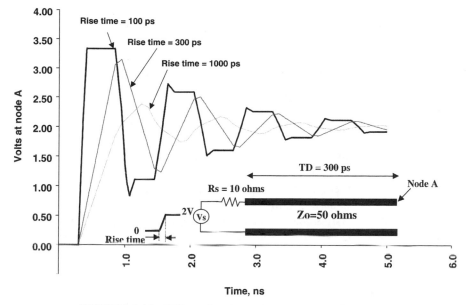

FIGURE 2.20 Effect of a slow edge rate: overdriven case.

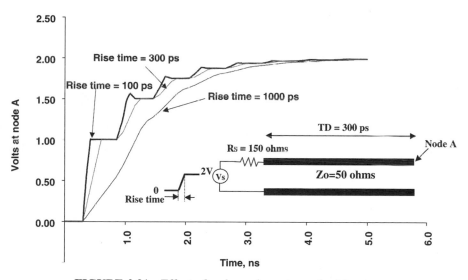

FIGURE 2.21 Effect of a slow edge rate: underdriven case.

wires and the lead frames of the chip packages are quite inductive. This makes it necessary to understand how these reactive elements affect the reflections in a system. In this section we introduce the effect that capacitors and inductors have on reflections. This knowledge will be used as a basis in future chapters, where capacitive and inductive parasitic effects are explored in more detail.

Reflections from a Capacitive Load. When a transmission line is terminated in a reactive element such as a capacitor, the waveforms at the driver and load will have a shape quite different from that of the typical transmission line response. Essentially, a capacitor is a time-dependent load, which will initially look like a short circuit when the signal reaches the capacitor and will look like an open circuit after the capacitor is fully charged. Let's consider the reflection coefficient at time = TD and at time = t_1. At time = TD, which is the time when the signal has propagated down the line and has reached the capacitive load, the capacitor will not be charged and will look like a short circuit. As described earlier in the chapter, a short circuit will have a reflection coefficient of -1. This means that the initial wave of magnitude V will be reflected off the load with a magnitude of $-V$, yielding an initial voltage of 0 V. The capacitor will then begin to charge at a rate dependent on τ, which is the time constant of an RC circuit, where C is the termination capacitor and R is the characteristic impedance of the transmission line. Once the capacitor is fully charged, the reflection coefficient will be 1 since the capacitor will resemble an open circuit. The voltage at the capacitor beginning at time $t = $ TD is governed by

$$V_{\text{capacitor}} = 2V_i(1 - e^{-(t-\text{TD})/\tau}), \qquad t > \text{TD} \tag{2.12}$$

$$\tau = CZ_o \tag{2.13}$$

Figure 2.22 shows a simulation of the response of a line terminated with a capacitive load. The load capacitance is 10 pF, the line length is 3.5 in. (TD = 500 ps), and the driver and transmission line impedance are both 50 Ω. Notice the shape of the waveform at the source (node A). It dips toward 0 at 1 ns, which is 2TD, the time that the reflection from the load arrives at the source. It dips toward zero because the initial reflection coefficient off the capacitor is -1, so the voltage reflected back toward the source is initially $V_i + (-V_i)$, where V_i is the initial voltage launched onto the transmission line. The capacitor then charges to a steady-state value of 2 V.

If the line is terminated with a parallel resistor and capacitor, as depicted in Figure 2.23, the voltage at the capacitor will depend on

$$V_{\text{capacitor}} = 2V_i \frac{R_L}{R_L + Z_o}(1 - e^{-(t-\text{TD})/\tau_1}), \qquad t > \text{TD} \tag{2.14}$$

and the time constant will depend on the C_L and the parallel combination of R_L and Z_o:

$$\tau_1 = \frac{C_L Z_o R_L}{R_L + Z_o} \tag{2.15}$$

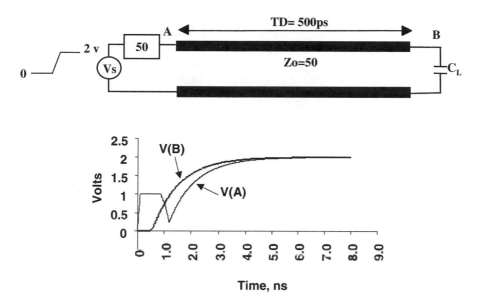

FIGURE 2.22 Transmission line terminated in a capacitive load.

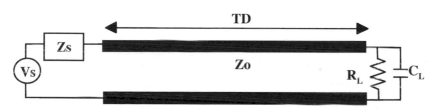

FIGURE 2.23 Transmission line terminated in a parallel capacitive and resistive load.

Reflection from an Inductive Load. When a series inductor appears in the electrical pathway on a transmission line, as depicted in Figure 2.24, it will also act as a time-dependent load. Initially, at time = 0, the inductor will resemble on open circuit. When a voltage step is applied initially, almost no current flows across the inductor. This produces a reflection coefficient of 1. The value of the inductor will determine how long the reflection coefficient will remain 1. If the inductor is large enough, the signal will double in magnitude. Eventually, the inductor will discharge its energy at a rate that depends on the time constant τ of an LR circuit, which will have a value of L/Z_o. Figure 2.25 shows the reflections from four different values of the series inductor depicted in Figure 2.24. Notice that the magnitude of the reflection and the decay time increased with increasing inductor value.

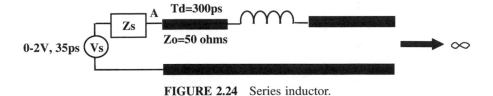

FIGURE 2.24   Series inductor.

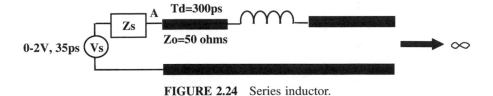

FIGURE 2.25   Reflection as seen at node *A* of Figure 2.24 for different inductor values.

2.4.5. Termination Schemes to Eliminate Reflections

As explained in later chapters, reflections on a transmission line can have a significant negative impact on the performance of a digital system. To minimize the negative impact of the reflections, methods must be developed to control them. Essentially, there are three ways to mitigate the negative impact of these reflections. The first method is to decrease the frequency of the system so that the reflections on the transmission line reach steady state before another signal is driven onto the line. This is usually impossible, however, for high-speed systems since it requires decreasing the operating frequency, producing a low-speed system. The second method is to shorten the PCB traces so that the reflections will reach steady state in a shorter time. This is usually not practical since doing so generally involves using a PCB board with a greater number of layers, which increases cost significantly. Additionally, shortening the traces may be physically impossible in some cases. These first two methods always have a limit at which the bus frequency increases to a point where the reflections won't reach steady state in one period. The third method is to terminate the transmission line with an impedance equal to the characteristic impedance of the line at either end of the transmission line and eliminate the reflections.

When the source end of the transmission line is designed to match the characteristic impedance of the transmission line, the bus is said to be *source terminated*. When a bus is source terminated, any reflections produced by a large impedance discontinuity at the far end of the line (such as an open circuit) are eliminated when they reach the source because the reflection coefficient will be zero. When a terminating resistor is placed at the far end of the line, the bus is said to be *parallel*, or *load terminated*. Multiple reflections will be eliminated at the load because the reflection coefficient at the load is zero. There are several different ways to implement these termination methodologies. There are advantages and disadvantages to each technique. Several techniques are summarized in the following sections.

On-Die Source Termination. On-die source termination requires that the *I–V* curve of the output buffer be very linear over the operating range and yield an *I–V* curve with an impedance very close to the transmission line impedance. Ideally, this is an optimum solution because it does not require any additional components that increase cost and consume area on the board. However, since there are numerous variables that drastically affect the output impedance of a buffer, it is difficult to achieve a good match between the buffer impedance and the line impedance. Some of the variables that affect the buffer impedance are silicon fabrication process variations, voltage, temperature, power delivery factors, and simultaneous switching noise. These variations will make it difficult to guarantee that the buffer impedance will match the line impedance. Figure 2.26 depicts this termination method.

Series Source Termination. Series source termination requires that a resistor be added in series with the output buffer. Figure 2.27 depicts the implementation of a series source termination. This type of termination requires that the sum of the buffer impedance and the value of the resistor be equal to the characteristic impedance of the line. This is usually best achieved by designing the *I–V* curve of the output buffer to yield a very low impedance such that the bulk of the impedance looking into the source will be contained in the resistor. Since precision resistors can be chosen, the effect of the on-die impedance variation due to process and environmental variations on the silicon that make on-die source termination difficult can be minimized. The total variation in impedance will be small because the resistor, rather than the output buffer itself, will comprise the bulk of the impedance. The disadvantages of this technique are that the resistors add cost to the board, and it consumes significant board area.

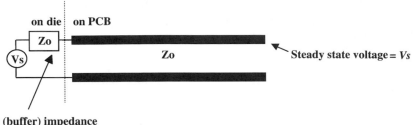

FIGURE 2.26   On-die source termination.

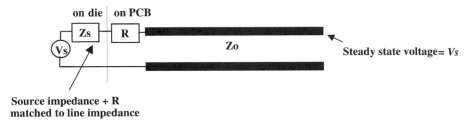

FIGURE 2.27   Series source termination.

Load Termination with a Resistive Load. Load or parallel termination with a resistive load eliminates the unknown variables associated with the buffer impedance because a precision resistor can be used. The reflections are eliminated at the load and low-impedance output buffers may be used. The disadvantage is that a large portion of the dc current will be shunted to ground, which exacerbates power delivery and thermal problems. The steady-state voltage will also be determined from the voltage divider between the source resistance and the load resistance, which creates the need for stronger buffers. Power delivery is a difficult problem to solve in modern computers. Laptops, for example, need very efficient power delivery systems since they require the use of batteries over prolonged periods of time. As power consumption increases, cost also increases, because more elaborate cooling mechanisms must be introduced to dissipate the excess heat. Figure 2.28 depicts this termination scheme.

AC Load Termination. Ac load termination uses a series capacitor and resistor at the load end of a transmission line to eliminate the reflections. The resistor R should be equal to the characteristic impedance of the transmission line, and the capacitor C_L should be chosen such that the RC time constant at the load is approximately equal to one or two rise times. It is advised that simulations be performed to choose the optimum capacitor value for the specific

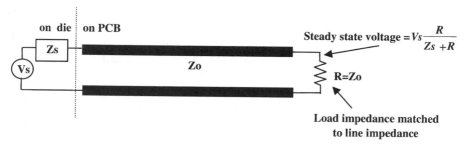

FIGURE 2.28 Load termination.

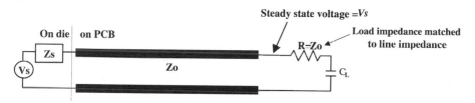

FIGURE 2.29 Ac load termination.

design. The premise behind this termination scheme is that the capacitor will initially act like a short circuit and the line will be terminated in its characteristic impedance by the resistor R for the duration of the rising or falling edge. The capacitor will then charge up and the steady-state voltage of the source, V_s, will be reached. The advantage of this technique is that the reflections are eliminated at the load with no dc power dissipation. The disadvantages are that the capacitive loading will increase the signal delay by slowing down the rising or falling times at the load. Furthermore, the additional resistors and capacitors consume board area and increase cost. Figure 2.29 depicts this termination scheme.

Common Termination Problems. One of the common obstacles encountered during bus design is that the characteristic impedance of the trace tends to vary significantly, due to PCB production variations. The PCB variation affects all the termination methods; however, it tends to have a bigger impact on source termination. Typical, low-cost PCB boards, for example, usually vary as much as $\pm 15\%$ from the target impedance over process. This means that if an engineer specifies a 65-Ω impedance for the lines on a PCB board, the vendor will guarantee that the impedance will be within 55.25 Ω (65 Ω − 15%) and 74.75 Ω (65 Ω + 15%). Finally, crosstalk will introduce additional variations in the impedance. The impact of the crosstalk-induced variations will depend on trace-to-trace spacing, the dielectric constant, and the cross-sectional geometry. Crosstalk is discussed thoroughly in Chapter 3.

For short lines, when the minimum digital pulse width is long compared to the time delay (TD) of the transmission line, source termination is desirable since it eliminates the need to shunt a portion of the driver current to ground. For long lines, where the width of the digital pulse is smaller than the time delay (TD) of the line, load termination is preferable. In the latter case, there will be multiple signals traveling down the transmission line at any given time (this is known as *pipeline mode*). Since reflections off the load will reflect back toward the source and interfere with the signals propagating down the line, the reflections must be eliminated at the load.

2.5. ADDITIONAL EXAMPLES

2.5.1. Problem

Assume that two components, U_1 and U_2, need to communicate with each other via a high-speed digital bus. The components are mounted on a standard four-layer motherboard with the stackup shown in Figure 2.30. The driving buffers on component U_1 have an impedance of 30 Ω, an edge rate of 100 ps, and a swing of 0 to 2 V. The traces on the PCB are required to be 50 Ω and 5 in. long. The relative dielectric constant of the board (ε_r) is 4.0, the transmission line is assumed to be a perfect conductor, and the receiver capacitance is small enough to be ignored. Figure 2.31 depicts the circuit topology.

2.5.2. Goals

1. Determine the correct cross-sectional geometry of the PCB shown in Figure 2.30 that will yield an impedance of 50Ω.
2. Calculate the time it takes for the signal to travel from the driver, U_1, to the receiver, U_2.

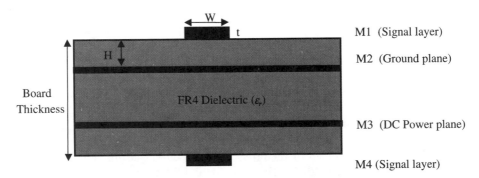

FIGURE 2.30   Standard four layer motherboard stackup.

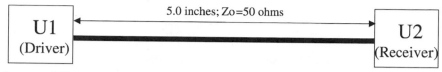

5.0 inches; Zo=50 ohms

U1
(Driver)

U2
(Receiver)

Edge rate = 100 ps
Output impedance, Rs=30 ohms

FIGURE 2.31   Topology of example circuits.

3. Determine the wave shape seen at U_2 when the system is driven by U_1.

4. Create an equivalent circuit of the system.

2.5.3. Calculating the Cross-Sectional Geometry of the PCB

Since the signal lines on the stackup depicted in Figure 2.30 are microstrips, the cross-sectional geometry can be determined from the equations in Figure 2.6. The fact that some of the microstrip lines are referencing a power plane instead of a ground plane should not cause confusion. For the purpose of transmission line design, a dc power plane will act like an ac ground and is treated as a ground plane in transmission line design. The consequences of this approximation are examined in subsequent chapters. Figure 2.6 shows that the trace impedance is a function of H, W, t, and ε_r (see Figure 2.30):

$$Z_{o_{microstrip}} = \frac{87}{\sqrt{\varepsilon_r + 1.41}} \ln \frac{5.98H}{0.8W + t}$$

The standard metal thickness of microstrip layers on a PCB board is usually on the order of 1.0 mil. Since the desired impedance, dielectric constant ε_r, and the thickness are known, this leaves one equation and two unknowns. Subsequently, one value of either H or W needs to be chosen. Typical PCB manufacturers will usually deliver a minimum trace width of 5 mils. Since small trace widths use up less board real estate, which allows the PCB to be physically smaller and less expensive, the minimum trace width of 5.0 mils is chosen. Plugging the known values of ε_r, W, t, and Z_o into the equation above and solving for H yields

$$50 = \frac{87}{\sqrt{4.0 + 1.41}} \ln \frac{5.98H}{0.8(5.0) + 1.0}$$

$$\frac{50\sqrt{5.41}}{87} = \ln \frac{5.98H}{5.0}$$

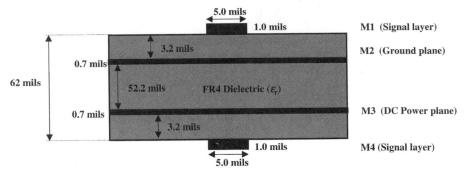

FIGURE 2.32   Resulting PCB stackup resulting in 50-ohm transmission lines.

$$e^{1.337} = 1.196H$$

$$H = 3.2 \text{ mils}$$

Figure 2.32 depicts the resulting stackup of the PCB.

Note the thickness of the inner metal and dielectric layers. The internal metal layers are only 0.7 mil thick instead of 1.0 mil thick. This is because the outer layers are usually tinplated to stop the outer copper traces, which are exposed to the environment, from oxidizing. The inner traces are not plated. Since a typical PCB board thickness requested by industry is 62 mils thick, the thickness of the dielectric layers between the power and ground plane is increased to achieve the desired board thickness.

2.5.4. Calculating the Propagation Delay

To calculate the time it takes for the signal to travel from U_1 to U_2, it is necessary to determine the propagation velocity for a signal traveling on the transmission line that was just designed. Since the transmission line is a microstrip, the effective dielectric constant must be used to calculate the propagation velocity. The effective dielectric constant is calculated using equations (2.6) and (2.7).

$$\varepsilon_e = \frac{\varepsilon_r + 1}{2} + \frac{\varepsilon_r - 1}{2}\left(1 + \frac{12H}{W}\right)^{-1/2} + F - 0.217(\varepsilon_r - 1)\frac{t}{\sqrt{WH}}$$

Since $W/H = 5/3.2 > 1.0$, then $F = 0$. Therefore,

$$\varepsilon_e = \frac{4.0 + 1}{2} + \frac{4.0 - 1}{2}\left[1 + \frac{12(3.2)}{5.0}\right]^{-1/2} + 0 - 0.217(4.0 - 1)\frac{1.0}{\sqrt{5.0(3.2)}}$$

$$= 2.84$$

The propagation velocity is determined using equation (2.2).

$$v = \frac{c}{\sqrt{\varepsilon_r}} = \frac{3.0 \times 10^8 \text{ m/s}}{\sqrt{2.84}} = 1.78 \times 10^8 \text{ m/s}$$

The time delay for the signal to propagate down the 5-in. line connecting U_1 and U_2 is then calculated using equation (2.4).

$$\text{TD} = \frac{\text{length}\sqrt{\varepsilon_r}}{c} = \frac{5.0 \text{ in.}}{1.78 \text{ m/s}} \left(\frac{0.0254 \text{ m}}{1.0 \text{ in.}} \right) = 713 \text{ ps}$$

2.5.5. Determining the Wave Shape Seen at the Receiver

The easiest way to calculate the wave shape seen at the receiver is to use a lattice (or bounce) diagram (Figure 2.33). It should be noted that a Bergeron diagram could also be used; however, since there is no nonlinear device at the receiver such as a diode, the lattice diagram is the preferred method of analysis. Refer to Section 2.4.

$$V_{\text{initial}} = V_{\text{in}} \frac{Z_{0_{\text{PCB}}}}{Z_{0_{\text{PCB}}} + R_s} = 2.0 \left(\frac{50}{50 + 30} \right) = 1.25 \text{ V}$$

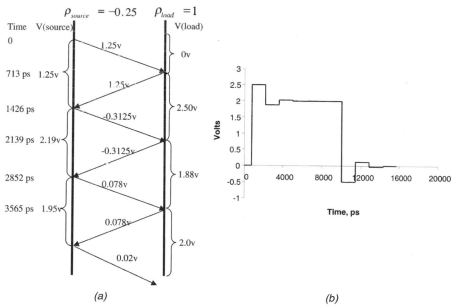

FIGURE 2.33 Determining the wave shape at the receiver: (a) lattice diagram; (b) waveform at U_2 when U_1 is driving.

$$\rho_{\text{source}} = \frac{R_s - Z_{o_{\text{PCB}}}}{R_s + Z_{o_{\text{PCB}}}} = \frac{30 - 50}{30 + 50} = -0.25$$

$$\rho_{\text{load}} = \frac{R_{\text{load}} - Z_{o_{\text{PCB}}}}{R_{\text{load}} + Z_{o_{\text{PCB}}}} = \frac{\infty - 50}{\infty + 50} = 1.0$$

where R_s is the buffer impedance of U_1 and R_{load} is the open circuit seen at U_2 (this assumes that the input capacitance to the buffer is very small, i.e., 2 to 3 pF).

2.5.6. Creating an Equivalent Circuit

To create the equivalent circuit model for this example, it is first necessary to determine the number of LC segments required. This is done using the rule of thumb presented in Section 2.3.3. To do so, it is convenient to convert the velocity calculated earlier to the units of inches/picosecond:

$$\text{velocity (in./ps)} = \frac{5.0 \text{ in.}}{713 \text{ ps}} = 0.0070 \text{ in./ps}$$

$$\text{segments} \geq 10 \left(\frac{\text{length}}{T_r v} \right) = 10 \left[\frac{5 \text{ in.}}{(100 \text{ ps})(0.0070 \text{ in./ps})} \right] = 71.4$$

Therefore, the minimum of 72 segments is required to create an accurate transmission line model.

Now the equivalent inductance and capacitance per unit inch is calculated using equations (2.1) and (2.5). Since this particular example is loss-free (i.e., a perfect conductor is assumed), equation (2.1) is reduced to $Z_o = \sqrt{L/C}$. The delay per unit inch is $TD = 713 \text{ ps}/5.0 \text{ in.} = 142.6 \text{ ps/in.} = \sqrt{LC}$, where L and C are per unit inch.

The equivalent L and C per unit inch are calculated by solving TD and Z_o for L and C.

$$C = \frac{TD}{Z_o} = \frac{\sqrt{LC}}{\sqrt{L/C}} = \frac{142.6 \text{ ps}}{50} = 2.85 \text{ pF}$$

$$L = (TD)(Z_o) = \sqrt{LC} \sqrt{\frac{L}{C}} = 7.130 \text{ nH}$$

Now, referring back to the rule of thumb presented in Section 2.3.3, the L and C per segment are calculated.

$$C_{\text{segment}} = \frac{(\text{length})(C/\text{in.})}{\text{segment}} = \frac{5.0 \text{ in.}(2.85 \text{ pF/in.})}{72} = 0.198 \text{ pF/segment}$$

$$L_{\text{segment}} = \frac{(\text{length})(L/\text{in.})}{segment} = \frac{5.0 \text{ in.}(7.13 \text{ nH/in.})}{72} = 0.495 \text{ nH/segment}$$

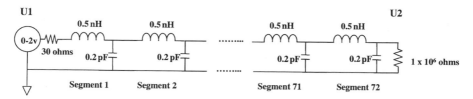

FIGURE 2.34 Final equivalent circuit.

To double check the calculations, the impedance and delay per segment can be calculated:

$$\text{delay/segment} = \frac{713 \text{ ps}}{72} = 9.9 \text{ ps} \approx \sqrt{(0.198 \text{ pF})(0.495 \text{ nH})} = 9.9 \text{ ps}$$

$$\text{impedance/segment} = \sqrt{\frac{0.495 \text{ nH}}{0.198 \text{ pF}}} = 50 \text{ }\Omega$$

The equivalent circuit of the transmission line is then constructed of 72 *LC* segments. The driving buffer at U_1 is depicted as a simple voltage source and a series resistor. The open circuit at the end of the line is approximated with a very large resistor (Figure 2.34).

CHAPTER 3

Crosstalk

Crosstalk, which is the coupling of energy from one line to another, will occur whenever the electromagnetic fields from different structures interact. In digital designs the occurrence of crosstalk is very widespread. Crosstalk will occur on the chip, on the PCB board, on the connectors, on the chip package, and on the connector cables. Furthermore, as technology and consumer demands push for physically smaller and faster products, the amount of crosstalk in digital systems is increasing dramatically. Subsequently, crosstalk will introduce significant hurdles into system design, and it is vital that the engineer learn the mechanisms that cause crosstalk and the design methodologies to compensate for it.

In multiconductor systems, excessive line-to-line coupling, or crosstalk, can cause two detrimental effects. First, crosstalk will change the performance of the transmission lines in a bus by modifying the effective characteristic impedance and propagation velocity, which will adversely affect system-level timings and the integrity of the signal. Additionally, crosstalk will induce noise onto other lines, which may further degrade the signal integrity and reduce noise margins. These aspects of crosstalk make system performance heavily dependent on data patterns, line-to-line spacing, and switching rates. In this section we introduce the mechanisms that cause crosstalk, present modeling methodology, and elaborate on the effects that crosstalk has on system-level performance.

3.1. MUTUAL INDUCTANCE AND MUTUAL CAPACITANCE

Mutual inductance is one of the two mechanisms that cause crosstalk. Mutual inductance L_m induces current from a driven line onto a quiet line by means of the magnetic field. Essentially, if the "victim" or quiet trace is in close enough proximity to the driven line such that its magnetic field encompasses the victim trace, a current will be induced on that line. The coupling of current via the magnetic field is represented in the circuit model by a mutual inductance.

The mutual inductance L_m will inject a voltage noise onto the victim proportional to the rate of change of the current on the driver line. The magnitude of this noise is calculated as

$$V_{\text{noise},L_m} = L_m \frac{dI_{\text{driver}}}{dt} \tag{3.1}$$

Since the induced noise is proportional to the rate of change, mutual inductance becomes very significant in high-speed digital applications.

Mutual capacitance is the other mechanism that causes crosstalk. Mutual capacitance is simply the coupling of two conductors via the electric field. The coupling due to the electric field is represented in the circuit model by a mutual capacitor. Mutual capacitance C_m will inject a current onto the victim line proportional to the rate in change of voltage on the driven line:

$$I_{\text{noise},C_m} = C_m \frac{dV_{\text{driver}}}{dt} \tag{3.2}$$

Again, since the induced noise is proportional to the rate of change, mutual capacitance also becomes very significant in high-speed digital applications.

It should be noted that equations (3.1) and (3.2) are only simple approximations used to explain the mechanisms of the coupled noise. Complete crosstalk formulas are presented later in the chapter.

3.2. INDUCTANCE AND CAPACITANCE MATRIX

In systems where significant coupling occurs between transmission lines, it is no longer adequate to represent the electrical characteristics of the line with just an inductance and a capacitance, as could be done with the single-transmission-line case presented in Chapter 2. It becomes necessary to consider the mutual capacitance and mutual inductance to fully evaluate the electrical performance of a transmission line in a multiconductor system. Equations (3.3) and (3.4) depict the typical method of representing the parasitics that govern the electrical performance of a coupled transmission line system. The inductance and capacitance matrices are known collectively as the *transmission line matrices*. The example shown is for an N-conductor system and is representative of what is usually reported by a field simulator (see Section 3.3), which is a tool used to calculate the inductance and capacitance matrices of a transmission line system.

$$\text{Inductance matrix} = \begin{bmatrix} L_{11} & L_{12} & \cdots & L_{1N} \\ L_{21} & L_{22} & & \\ \vdots & & \ddots & \\ L_{N1} & & & L_{NN} \end{bmatrix} \tag{3.3}$$

where L_{NN} is the self-inductance of line N and L_{MN} is the mutual inductance between lines M and N.

$$\text{Capacitance matrix} = \begin{bmatrix} C_{11} & C_{12} & \cdots & C_{1N} \\ C_{21} & C_{22} & & \\ \vdots & & \ddots & \\ C_{N1} & & & C_{NN} \end{bmatrix} \tag{3.4}$$

where C_{NN} is the total capacitance seen by line N, which consists of conductor N's capacitance to ground plus all the mutual capacitance to other lines and C_{NM} = mutual capacitance between conductors N and M.

Example 3.1: Two-Conductor Transmission Line Matrices. The capacitance matrix for Figure 3.1 is

$$\text{capacitance matrix} = \begin{bmatrix} C_{11} & C_{12} \\ C_{21} & C_{22} \end{bmatrix} \tag{3.5}$$

where the self-capacitance of line 1, C_{11}, is the sum of the capacitance of line 1 to ground (C_{1g}) plus the mutual capacitance from line 1 to line 2 (C_{12}):

$$C_{11} = C_{1g} + C_{12} \tag{3.6}$$

Additionally, the inductance matrix for the system depicted in Figure 3.1 is

$$\text{inductance matrix} = \begin{bmatrix} L_{11} & L_{12} \\ L_{21} & L_{22} \end{bmatrix} \tag{3.7}$$

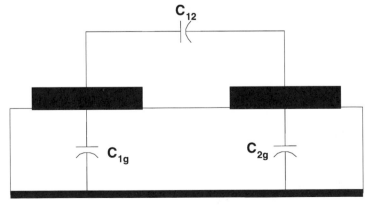

Ground plane

FIGURE 3.1 Simple two-conductor system used to explain the parasitic matrices.

where L_{11} is the self-inductance for line 1 and L_{12} is the mutual inductance between lines 1 and 2. L_{11} is not the sum of the self-inductance and the mutual inductance as in the case of C_{11}.

3.3. FIELD SIMULATORS

Field simulators are used to model the electromagnetic interaction between transmission lines in a multiconductor system and to calculate trace impedances, propagation velocity, and all self and mutual parasitics. The outputs are typically matrices that represent the effective inductance and capacitance values of the conductors. These matrices are the basis of all equivalent circuit models and are used to calculate characteristic impedance, propagation velocity, and crosstalk. Field simulators fall into two general categories: *two-dimensional*, or *electrostatic*, and *three-dimensional*, or *full-wave*. Most two-dimensional simulators will give the inductance and capacitance matrices as a function of conductor length, which is usually most suitable for interconnect analysis and modeling. The advantage of the two-dimensional or electrostatic simulators is that they are very easy to use and typically take a very short amount of time to complete the calculations. The disadvantages of two-dimensional simulators are that they will simulate only relatively simple geometries, they are based on static calculations of the electric field, and they usually do not calculate frequency-dependent effects such as internal inductance or skin effect resistance. This is usually not a significant obstacle, however, since interconnects are usually simple structures and there are alternative methods of calculating effects such as frequency-dependent resistance and inductance.

Additionally, there are several three-dimensional, or full-wave simulators on the market. The advantage of a full-wave simulator is that they will simulate complex three-dimensional geometries, and they will predict frequency-dependent losses, internal inductance, dispersion, and most other electromagnetic phenomena, including radiation. The simulators essentially solve Maxwell's equations directly for an arbitrary geometry. The disadvantage is that they are very difficult to use, and simulations typically take hours or days rather than seconds. Additionally, the output from a full wave simulator is often in the form of S parameters, which are not very useful for interconnect simulations for digital applications. For the purposes of this book, emphasis is placed on electrostatic two-dimensional simulators.

3.4. CROSSTALK-INDUCED NOISE

As explained in Section 3.1, crosstalk is the result of mutual capacitance C_m in conjunction with mutual inductance L_m between adjacent conductors. The magnitude of the noise induced onto the adjacent transmission lines will de-

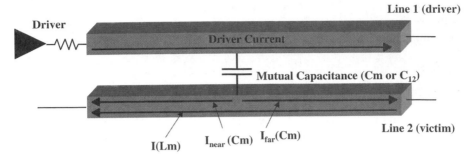

FIGURE 3.2 Crosstalk-induced current trends caused by mutual inductance and mutual capacitance.

pend on the magnitude of both the mutual inductance and the mutual capacitance. For example, if a signal is injected into line 1 of Figure 3.2, current will be generated on the adjacent line by means of L_m and C_m. For simplicity, let us define a few terms. *Near-end crosstalk* is defined as the crosstalk seen on the victim line at the end closest to the driver (this is sometimes called *backward crosstalk*). *Far-end crosstalk* refers to the crosstalk observed on the victim line farthest away from the driver (sometimes known as *forward crosstalk*). The current generated on the victim line due to mutual capacitance will split and flow toward both ends of the adjacent line. The current induced onto the victim line as a result of the mutual inductance will flow from the far end toward the near end of the victim line since mutual inductance creates current flow in the opposite direction. As a result, the crosstalk currents flowing toward the near and far ends can be broken down into several components (see Figure 3.2):

$$I_{\text{near}} = I(L_m) + I_{\text{near}}(C_m) \tag{3.8}$$

$$I_{\text{far}} = I_{\text{far}}(C_m) - I(L_m) \tag{3.9}$$

The shape of the crosstalk noise seen at the near and far ends of the victim line can be deduced by looking at Figure 3.3. As a digital pulse travels down a transmission line, the rising and falling edges will induce noise continuously on the adjacent line. For this discussion, assume that the rise and fall times are much smaller than the delay of the line. As described previously, a portion of the crosstalk noise will travel toward the near end of the line and a portion will travel toward the far end. The portions traveling toward the near and far ends will be referred to as the near- and far-end crosstalk pulses, respectively. As depicted by Figure 3.3, the far-end crosstalk pulse will travel concurrently with the edge of the signal on the driving line. The near-end crosstalk pulse will originate at the edge and propagate back toward the near end. Subsequently, when the signal edge reaches the far end of the driving line at time $t = TD$ (where TD is the electrical delay of the transmission line), the driving signal

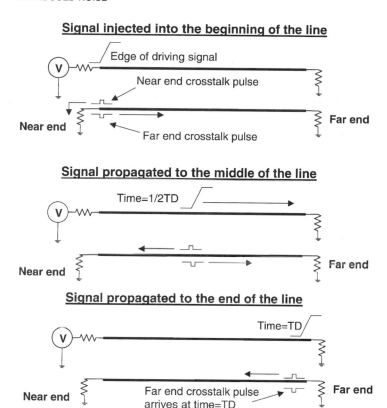

FIGURE 3.3 Graphical representation of crosstalk noise.

and the far-end crosstalk will be terminated by the resistor. The last portion of the near-end crosstalk induced on the victim line just prior to the signal being terminated, however, will not arrive at the near end until time $t = 2TD$ because it must propagate the entire length of the line to return. Therefore, for a pair of terminated transmission lines, the near-end crosstalk will begin at time $t = 0$ and have a duration of 2TD, or twice the electrical length of the line. Furthermore, the far-end crosstalk will occur at time $t = TD$ and have a duration approximately equal to the signal rise or fall time.

The magnitude and shape of the crosstalk noise depend heavily on the amount of coupling and the termination. The equations and illustrations in Figure 3.4 describe the maximum voltage levels induced on a quiet victim line as a result of crosstalk, assuming a variety of victim line termination strategies [DeFalco, 1970]. The driving line is terminated to eliminate complications due to multiple reflections. These equations should be used primarily to estimate the magnitude of crosstalk noise and to understand the impact of a particular termination strategy. For topologies more complicated then the ones shown in Figure 3.4, a simulator such as SPICE is required.

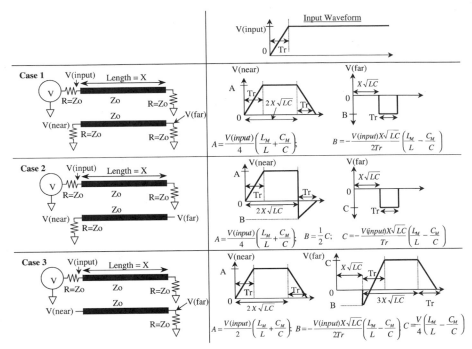

FIGURE 3.4 Digital crosstalk noise as a function of victim termination.

The equations in Figure 3.4 assume that the line delay TD is at least twice the rise time:

$$TD = X\sqrt{LC} \tag{3.10}$$

where X is the line length and L and C are the self-inductance and capacitance of the transmission line per unit length. Note that if $T_r > 2X\sqrt{LC}$ (i.e., the edge rate is greater than twice the line delay), the near-end crosstalk will fail to achieve its maximum amplitude. To calculate the correct crosstalk voltages when $T_r > 2X\sqrt{LC}$, simply multiply the near-end crosstalk by $2X\sqrt{LC}/T_r$. The far-end crosstalk equations do not need to be adjusted.

Note that the near-end crosstalk is independent of the input rise time when the rise time is short compared to the line delay (long-line case) and is dependent on the rise time when the rise time is long compared to the line length (short-line case). For this subject matter, the definition of a long line is when the electrical delay of the line is at least one-half the signal rise time (or fall time). Moreover, the near-end magnitude is independent of length for the long-line case, while the far end always depends on rise time and length.

It should be noted that the formulas in Figure 3.4 assume that the termination resistors on the victim line are matched to the transmission line, which eliminates any effects of imperfect terminations. To compensate for this, simply use the transmission line reflection concepts introduced in Chapter 2. For example, assume that the termination R in case 1 of Figure 3.4 is not equal to the characteristic impedance of the victim transmission line (for simplicity, it will still be assumed that the termination on the driver line remains perfectly matched to its characteristic impedance). In this case the near- and far-end reflections must be added to the respective crosstalk voltages. The resultant crosstalk signal, accounting for nonideal terminations, is calculated as

$$V_x = V_{\text{crosstalk}} \left(1 + \frac{R - Z_o}{R + Z_o}\right) \tag{3.11}$$

where V_x is the crosstalk at the near or far end of the victim line adjusted for a nonperfect termination, R the impedance of the termination, Z_o the characteristic impedance of the transmission line, and $V_{\text{crosstalk}}$ the value calculated from Figure 3.4.

POINTS TO REMEMBER

- If the rise or fall time is short compared to the delay of the line, the near-end crosstalk noise is independent of the rise time.
- If the rise or fall time is long compared to the delay of the line, the near-end crosstalk noise is dependent on the rise time.
- The far-end crosstalk is always dependent on the rise or fall time.

Example 3.2: Calculating Crosstalk-Induced Noise for a Matched Terminated System. Assume the two-conductor system shown in Figure 3.5, where $Z_o \approx$ 70 Ω, the termination resistors = 70 Ω, $V(\text{input}) = 1.0$ V, $T_r = 100$ ps, and $X = 2$ in. Determine the near- and far-end crosstalk magnitudes assuming the following capacitance and inductance matrices:

$$L/\text{in.} = \begin{bmatrix} 9.869 \text{ nH} & 2.103 \text{ nH} \\ 2.103 \text{ nH} & 9.869 \text{ nH} \end{bmatrix}$$

$$C/\text{in.} = \begin{bmatrix} 2.051 \text{ pF} & 0.239 \text{ pF} \\ 0.239 \text{ pF} & 2.051 \text{ pF} \end{bmatrix}$$

SOLUTION:

$$V(1) = \frac{V(\text{input})}{4}\left(\frac{L_{12}}{L_{11}} + \frac{C_{12}}{C_{11}}\right) = \frac{1}{4}\left(\frac{2.103 \text{ nH}}{9.869 \text{ nH}} + \frac{0.239 \text{ pF}}{2.051 \text{ pF}}\right) = 0.082 \text{ V}$$

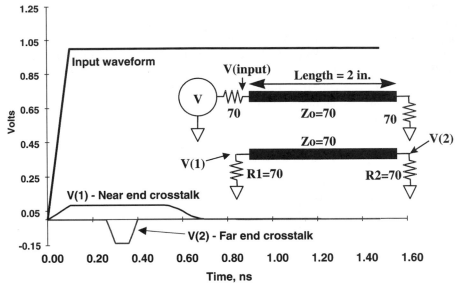

FIGURE 3.5 Near- and far-end crosstalk pulses.

$$V(2) = -\frac{V(\text{input})(X\sqrt{LC})}{2T_r}\left(\frac{L_{12}}{L_{11}} - \frac{C_{12}}{C_{11}}\right)$$

$$= \frac{1[2\sqrt{(9.869 \text{ nH})(2.051 \text{ pF})}]}{2(100 \text{ ps})}\left(\frac{2.103 \text{ nH}}{9.869 \text{ nH}} - \frac{0.239 \text{ pF}}{2.051 \text{ pF}}\right) = -0.137 \text{ V}$$

Figure 3.5 also shows a simulation of the system in this example. It can readily be seen that the near- and far-end crosstalk-induced noise matches the equations of Figure 3.4.

Example 3.3: Calculating Crosstalk-Induced Noise for a Nonmatched Terminated System. Consider the same two-conductor system of Example 3.2. If $R_1 = 45$ and $R_2 = 100$ Ω, what are the respective near- and far-end crosstalk voltages?

SOLUTION:

$$Z_o = \sqrt{\frac{L_{11}}{C_{11}}} = \sqrt{\frac{9.869 \text{ nH}}{2.051 \text{ pF}}} = 69.4 \text{ } \Omega$$

$$V(1) = V_{\text{crosstalk}}\left(1 + \frac{R - Z_o}{R + Z_o}\right) = 0.082\left(1 + \frac{45 - 69.4}{45 + 69.4}\right) = 0.0645 \text{ V}$$

$$V(2) = V_{\text{crosstalk}}\left(1 + \frac{R - Z_o}{R + Z_o}\right) = -0.137\left(1 + \frac{100 - 69.4}{100 + 69.4}\right) = -0.162 \text{ V}$$

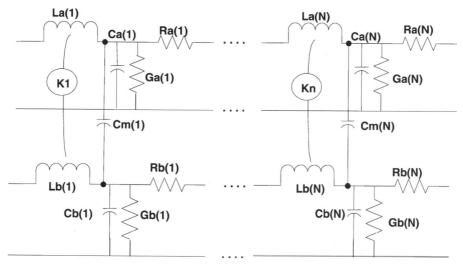

FIGURE 3.6 Equivalent circuit model of two coupled lines.

3.5. SIMULATING CROSSTALK USING EQUIVALENT CIRCUIT MODELS

Equivalent circuit models are the most general method of simulating crosstalk. Figure 3.6 depicts an N-segment equivalent circuit model of two coupled lines as modeled in SPICE, where N is the number of sections required such that the model will behave as a continuous transmission line and not as a series of lumped inductors, capacitors, and resistors. As mentioned in Chapter 2, the number of segments, N, in a transmission line model depends on the fastest edge rate used in the simulation. A good rule of thumb is that the propagation delay of a single segment should be less than or equal to one-tenth of the rise time (see Chapter 2 for a full explanation).

The mutual inductance is typically modeled in SPICE-type simulators with a coupling factor K:

$$K = \frac{L_{12}}{\sqrt{L_{11}L_{22}}} \tag{3.12}$$

where L_{12} is the mutual inductance between lines 1 and 2, and L_{11} and L_{22} are the self-inductances of lines 1 and 2, respectively.

Example 3.4: Creating a Coupled Transmission Line Model. Assume that a pair of coupled transmission lines is 5 in. long and a digital signal with a rise time of 100 ps is to be simulated. Given the following inductance and capacitance matrices, calculate the characteristic impedance, the total propagation delay, the inductive coupling factor, the number of required segments, the maximum delay per segment, and the maximum L, R, C, G, C_m, and K values for one segment.

$$\text{Capacitance matrix (per unit inch)} = \begin{bmatrix} 2 \text{ pF} & 0.1 \text{ pF} \\ 0.1 \text{ pF} & 2 \text{ pF} \end{bmatrix}$$

$$\text{Inductance matrix (per unit inch)} = \begin{bmatrix} 9 \text{ nH} & 0.7 \text{ nH} \\ 0.7 \text{ nH} & 9 \text{ nH} \end{bmatrix}$$

SOLUTION: Characteristic impedance:

$$Z_o = \sqrt{\frac{L_{11}}{C_{11}}} = \sqrt{\frac{9 \text{ nH}}{2 \text{ pF}}} = 67.09$$

Total propagation delay:

$$TD = \sqrt{L_{11}C_{11}} = \sqrt{(9 \text{ nH})(2 \text{ pF})} = 134 \text{ ps/in.} \rightarrow \times 5 \text{ in.} = 670 \text{ ps}$$

Inductive coupling factor:

$$K = \frac{L_{12}}{\sqrt{L_{11}L_{22}}} = \frac{0.7 \text{ nH}}{9 \text{ nH}} = 0.078$$

Minimum number of segments (see Chapter 2):

$$N = \text{segments} = 10\frac{X}{vT_r} = 10\frac{(5 \text{ in.})(134 \text{ ps/in.})}{100 \text{ ps}} = 67$$

The transmission line is 5 in. long and there need to be a minimum of 67 total segments for the model to behave like a transmission line. Subsequently, since the capacitance and inductance matrices are reported per unit inch, the L, C, and C_m values must be multiplied by $\frac{5}{67}$ (in./segment). Subsequently, $L(N) = 0.67$ nH, $C_g(N) = C_{11} - C_{12} = 0.1425$ pF, and $C_m(N) = 0.0075$ pF. Since the inductive coupling factor K has no units [as calculated with equation (3.12) all the units cancel], it is not necessary to scale it for each segment. Therefore, $K(N) = 0.078$.

3.6. CROSSTALK-INDUCED FLIGHT TIME AND SIGNAL INTEGRITY VARIATIONS

The effective characteristic impedance and propagation delay of a transmission line will change with different switching patterns in systems where significant coupling between traces exists. Electric and magnetic fields interact in specific ways that depend on the data patterns on the coupled traces. These specific patterns can effectively reduce or increase the effective parasitic inductance and capacitance seen by a single line. Since the transmission line parameters vary with data patterns and are essential for accurate timing and signal integrity analysis, it is important to consider these variations in high-trace-density and high-speed systems. In this section we explain the effect that crosstalk-induced impedance and velocity changes will have on signal integrity and timings. Additionally, simplified signal integrity analysis methodologies, which correctly model multiple conductors in a computer bus, will be introduced.

3.6.1. Effect of Switching Patterns on Transmission Line Performance

When multiple traces are in close proximity, the electric and magnetic fields will react with each other in specific ways that depend on the signal patterns present on the lines. The significance of this is that these interactions will alter the effective characteristic impedance and the velocity of the transmission lines. This phenomenon is magnified when many transmission lines are switching in close proximity to one another, and it effectively causes a bus to exhibit pattern-dependent impedance and velocity behavior. This can significantly affect the performance of a bus, so it is imperative to consider it in system design.

Odd Mode. Odd-mode propagation mode occurs when two coupled transmission lines are driven with equal magnitude and 180° out of phase with one another. The effective mutual capacitance of the transmission line will increase by twice the mutual capacitance, and the equivalent inductance will decrease by the mutual inductance. To examine the effect that odd-mode propagation on two adjacent traces will have on the characteristic impedance and on the velocity, consider Figure 3.7.

In odd-mode propagation, the currents on the lines, I_1 and I_2, will always be driven with equal magnitude but in opposite directions. First, let's consider the effect of mutual inductance. Refer to Figure 3.8. Assume that $L_{11} = L_{22} = L_0$. Remember that the voltage produced by the inductive coupling is calculated by equation (3.1). Subsequently, applying Kirchhoff's voltage law produces

$$V_1 = L_0 \frac{dI_1}{dt} + L_m \frac{dI_2}{dt} \tag{3.13}$$

$$V_2 = L_0 \frac{dI_2}{dt} + L_m \frac{dI_1}{dt} \tag{3.14}$$

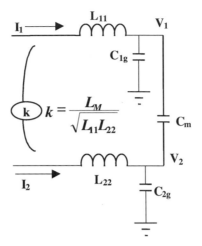

FIGURE 3.7 Equivalent circuit model used to derive the impedance and velocity variations for odd- and even-mode switching patterns.

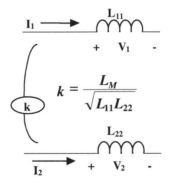

FIGURE 3.8 Simplified circuit for determining the equivalent odd-mode inductance.

Since the signals for odd-mode switching are always opposite, it is necessary to substitute $I_1 = -I_2$ and $V_1 = -V_2$ into (3.13) and (3.14). This yields

$$V_1 = L_0 \frac{dI_1}{dt} + L_m \frac{d(-I_1)}{dt} = (L_0 - L_m)\frac{dI_1}{dt} \tag{3.15}$$

$$V_2 = L_0 \frac{dI_2}{dt} + L_m \frac{d(-I_2)}{dt} = (L_0 - L_m)\frac{dI_2}{dt} \tag{3.16}$$

Therefore, the equivalent inductance seen by line 1 in a pair of coupled transmission lines propagating in odd mode is

$$L_{\text{odd}} = L_{11} - L_m = L_{11} - L_{12} \tag{3.17}$$

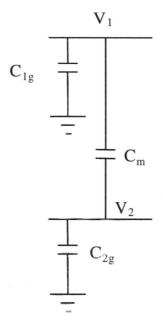

FIGURE 3.9 Simplified circuit for determining the equivalent odd mode capacitance.

Similarly, the effect of the mutual capacitance can be derived. Refer to Figure 3.9. Applying Kirchhoff's current law at nodes V_1 and V_2 yields (assume that $C_{1g} = C_{2g} = C_0$)

$$I_1 = C_0\frac{dV_1}{dt} + C_m\frac{d(V_1 - V_2)}{dt} = (C_0 + C_m)\frac{dV_1}{dt} - C_m\frac{dV_2}{dt} \qquad (3.18)$$

$$I_2 = C_0\frac{dV_2}{dt} + C_m\frac{d(V_2 - V_1)}{dt} = (C_0 + C_m)\frac{dV_2}{dt} - C_m\frac{dV_1}{dt} \qquad (3.19)$$

Again, substitution of $I_1 = -I_2$ and $V_1 = -V_2$ for odd-mode propagation yields

$$I_1 = C_0\frac{dV_1}{dt} + C_m\frac{d(V_1 - (-V_1))}{dt} = (C_{1g} + 2C_m)\frac{dV_1}{dt} \qquad (3.20)$$

$$I_2 = C_0\frac{dV_2}{dt} + C_m\frac{d(V_2 - (-V_2))}{dt} = (C_{2g} + 2C_m)\frac{dV_2}{dt} \qquad (3.21)$$

Therefore, the equivalent capacitance seen by trace 1 in a pair of coupled transmission line propagating in odd mode is

$$C_{odd} = C_{1g} + 2C_m = C_{11} + C_m \qquad (3.22)$$

Subsequently, the equivalent impedance and delay for a coupled pair of transmission lines propagating in an odd-mode pattern are

$$Z_{odd} = \sqrt{\frac{L_{odd}}{C_{odd}}} = \sqrt{\frac{L_{11} - L_{12}}{C_{11} + C_{12}}} \qquad (3.23)$$

$$TD_{odd} = \sqrt{L_{odd}C_{odd}} = \sqrt{(L_{11} - L_{12})(C_{11} + C_{12})} \qquad (3.24)$$

Even Mode. Even-mode propagation mode occurs when two coupled transmission lines are driven with equal magnitude and are in phase with one another. The effective capacitance of the transmission line will decrease by the mutual capacitance and the equivalent inductance will increase by the mutual inductance. To examine the effect that even-mode propagation on two adjacent traces will have on the characteristic impedance and on the velocity, consider Figure 3.7.

In even-mode propagation, the currents on the lines, I_1 and I_2, will always be driven with equal magnitude and in the same direction. First, let's consider the effect of mutual inductance. Again, refer to Figure 3.8. The analysis that was done for odd-mode switching can be done to determine the effective even-mode capacitance and inductance. For even-mode propagation, $I_1 = I_2$ and $V_1 = V_2$; therefore, equations (3.13) and (3.14) yield

$$V_1 = L_0 \frac{dI_1}{dt} + L_m \frac{d(I_1)}{dt} = (L_0 + L_m) \frac{dI_1}{dt} \qquad (3.25)$$

$$V_2 = L_0 \frac{dI_2}{dt} + L_m \frac{d(I_2)}{dt} = (L_0 + L_m) \frac{dI_2}{dt} \qquad (3.26)$$

Therefore, the equivalent inductance seen by line 1 in a pair of coupled transmission line propagating in even mode is

$$L_{even} = L_{11} + L_m \qquad (3.27)$$

Similarly, the effect of the mutual capacitance can be derived. Again, refer to Figure 3.9.

Substituting $I_1 = I_2$ and $V_1 = V_2$ for even-mode propagation, therefore, equations (3.18) and (3.19) yield

$$I_1 = C_0 \frac{dV_1}{dt} + C_m \frac{d(V_1 - V_1)}{dt} = (C_0) \frac{dV_1}{dt} \qquad (3.28)$$

$$I_2 = C_0 \frac{dV_2}{dt} + C_m \frac{d(V_2 - V_2)}{dt} = (C_0) \frac{dV_2}{dt} \qquad (3.29)$$

Therefore, the equivalent capacitance seen by trace 1 in a pair of coupled transmission line propagating in even mode is

$$C_{\text{even}} = C_0 = C_{11} - C_m \qquad (3.30)$$

Subsequently, the even-mode transmission characteristics for a coupled two-line system are

$$Z_{\text{even}} = \sqrt{\frac{L_{\text{even}}}{C_{\text{even}}}} = \sqrt{\frac{L_{11} + L_{12}}{C_{11} - C_{12}}} \qquad (3.31)$$

$$\text{TD}_{\text{even}} = \sqrt{L_{\text{even}} C_{\text{even}}} = \sqrt{(L_{11} + L_{12})(C_{11} - C_{12})} \qquad (3.32)$$

Figure 3.10 depicts the odd- and even-mode electric and magnetic field patterns for a simple two-conductor system. As explained in Chapter 2, the magnetic field lines will always be perpendicular to the electric field lines (assuming TEM mode). Notice that both conductors are at the same potential for even-mode propagation. Since there is no voltage differential, there can be no effect of capacitance between the lines. This is an easy way to remember that the mutual capacitance is subtracted from the total capacitance for the even mode. Since the conductors are always at different potentials in odd-mode propagation, an effect of the capacitance between the two lines must exist. This is an easy way to remember that you add the mutual capacitance for odd mode.

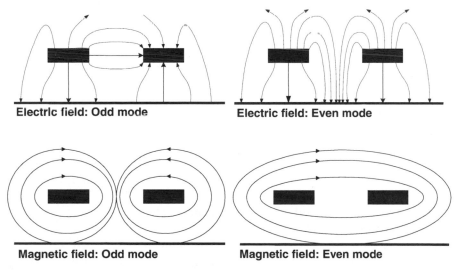

Electric field: Odd mode **Electric field: Even mode**

Magnetic field: Odd mode **Magnetic field: Even mode**

FIGURE 3.10 Odd- and even-mode electric and magnetic field patterns for a simple two-conductor system.

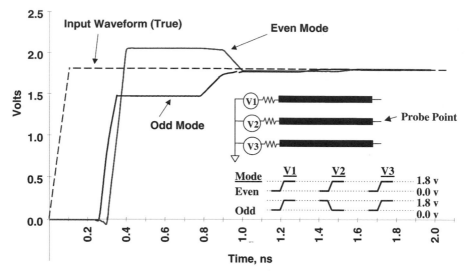

FIGURE 3.11   Effect of switching patterns on a three-conductor system.

Figure 3.11 demonstrates how odd- and even-mode impedance and velocity changes will affect the signals propagating on a transmission line. Although there are only three conductors in this example, the simulation shows the effect that crosstalk will have on a given net, which is routed between other nets in a computer bus. The figure shows that signal integrity and velocity (assuming a microstrip) can be affected dramatically by the crosstalk and is a function of switching patterns. Since signal integrity depends directly on the source (buffer) and transmission line impedance, the degree of coupling and the switching pattern will significantly affect system performance, due to the fact that the effective characteristic impedance of the traces will change. Moreover, if the transmission line is a microstrip, the velocity will change, which will affect timings, especially for long lines. This type of analysis using three conductors is usually adequate for design purposes and is a good first-order approximation of the true system performance because the nearest-neighbor lines have the most significant effect.

It is interesting to note that the mutual inductance is always added or subtracted in the opposite manner as the mutual capacitance for odd- and even-mode propagation. This can be visualized by looking at the fields in Figure 3.10. For example, in odd-mode propagation, the effect of the mutual capacitance must be added because the conductors are at different potentials. Additionally, since the currents in the two conductors are always flowing in opposite directions, the currents induced on each line due to the coupling of the magnetic fields always oppose each other and cancel out any effect due to the mutual inductance. Subsequently, the mutual inductance must be subtracted and the mutual capacitance must be added to calculate the odd-mode characteristics.

These characteristics of even- and odd-mode propagation are due to the assumption that the signals are propagating only in TEM (transverse electromagnetic) mode, which means that the electric and magnetic fields are always orthogonal to each other. As a result of the TEM assumption, the product of L and C remain constant for homogeneous systems (where the fields are contained within a single dielectric). Thus, in a multiconductor homogeneous system, such as a stripline array, if L is increased by the mutual inductance, C must be decreased by the mutual capacitance such that LC remains constant. Subsequently, a stripline or buried microstrip which is embedded in a homogeneous dielectric should not exhibit velocity variations due to different switching modes. It will, however, exhibit pattern-dependent impedance differences.

In a nonhomogeneous system (where the electric fields will fringe through more than one dielectric material) such as an array of microstrip lines, LC is not held constant for different propagation modes because the electromagnetic fields are traveling partially in air and partially in the dielectric material of the board. In a microstrip system the effective dielectric constant is a weighted average between air and the dielectric material of the board. Because the field patterns change with different propagation modes, the effective dielectric constant will change depending on the field densities contained within the board dielectric material and the air. Thus the LC product will be mode dependent in a nonhomogeneous system. The LC product will, however, remain constant for a given mode. Subsequently, a microstrip will exhibit both a velocity and impedance change, due to different switching patterns.

POINTS TO REMEMBER

- Odd-mode impedance will always be lower than the single-line case.
- Even-mode impedance will always be higher than the single-line case.
- There are no velocity variations due to crosstalk in a stripline.
- Crosstalk will induce velocity variations in a microstrip.

3.6.2. Simulating Traces in a Multiconductor System Using a Single-Line Equivalent Model

Two coupled conductors can be modeled as one conductor by determining the effective odd- or even-mode impedance and propagation delay of the transmission line pair and substituting these parameters into a single-line model. This technique can be expanded to determine the effective crosstalk-induced impedance and delay variations for a multiconductor system. This will allow the designer to estimate the worst-case effects of crosstalk during a bus design prior to an actual layout. This greatly increases the efficiency of the prelayout design because it is significantly less computationally intense than simulating with equivalent circuit models, and it allows crosstalk-induced impedance and velocity changes to be accounted for early in the design process.

To utilize this technique in multiconductor simulations, it is necessary to introduce a methodology to simulate one conductor in place of several conductors. This method is useful for quickly approximating the variations in impedance and delay that a line will experience due to crosstalk from several lines. Initially, the equivalent capacitance and inductance seen by the target line must be determined for a specific switching pattern. This can be accomplished by observing pairs of lines in a multiconductor system and determining the equivalent odd- or even-mode capacitance and inductance values. The values from the pairs can then be combined to produce the final value. To do so, the mutual components are either added or subtracted from the total capacitance or self-inductance of the target line. When the signal line is switching in phase with the target line, the mutual capacitance between the two lines is subtracted from the total capacitance of the target line, and the mutual inductance is added. Conversely, when the signal line is switching out of phase with the target line, the mutual capacitance between those two lines is added to the total capacitance of the target line and the mutual inductance is subtracted. The effective capacitance and inductance are used to calculate the equivalent characteristic impedance and propagation velocity for the single-line equivalent model (SLEM). Assume, for example, that the switching pattern of the three-conductor system shown in Figure 3.12a is such that all bits on the net are switching in phase.

If the primary goal is to evaluate this multiconductor system (Figure 3.12a) quickly with a single line, and the target is line 2, the equivalent impedance and time delay per unit length seen by line 2 can be calculated as

$$Z_{2,\text{eff}} = \sqrt{\frac{L_{22} + L_{12} + L_{23}}{C_{22} - C_{12} - C_{23}}} \tag{3.33}$$

$$\text{TD}_{2,\text{eff}} = \sqrt{(L_{22} + L_{12} + L_{23})(C_{22} - C_{12} - C_{23})} \tag{3.34}$$

Example 3.5: Pattern-Dependent Impedance and Delay. Assume that the switching pattern of a three-conductor system is such that bits 1 and 2 are switching in phase and bit 3 is switching 180° out of phase, as shown in Figure 3.12b (line 2 is still the victim).

Even-mode capacitance of conductor 2 with $1 = C_{22} - C_{12}$
Odd-mode capacitance of conductor 2 with $3 = C_{22} + C_{23}$

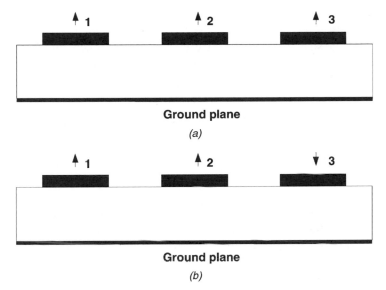

FIGURE 3.12   Example switching pattern: (*a*) all bits switching in phase; (*b*) bits 1 and 2 switching in phase, bit 3 switching 180° out of phase.

Equivalent capacitance of conductor $2 = C_{22} - C_{12} + C_{23}$
Even-mode inductance of conductor 2 with $1 = L_{22} + L_{12}$
Odd-mode inductance of conductor 2 with $3 = L_{22} - L_{23}$
Equivalent inductance of conductor $2 = L_{22} + L_{12} - L_{23}$

The equivalent single-conductor model of this multiconductor system would have the following transmission line parameters:

$$Z_{2,\text{eff}} = \sqrt{\frac{L_{22} + L_{12} - L_{23}}{C_{22} - C_{12} + C_{23}}}$$

$$\text{TD}_{2,\text{eff}} = \sqrt{(L_{22} + L_{12} - L_{23})(C_{22} - C_{12} + C_{23})}$$

It should be noted that this simplified technique is best used during the design phase of a bus when buffer impedances and line-to-line spacing are being chosen. The reader should also be reminded that this technique is useful only for signals traveling in the same direction. For crosstalk effects of signals traveling in opposite directions, equivalent circuit models and a circuit simulator must be utilized.

It should also be noted that the SLEM method is only an approximation that is useful for quickly narrowing down the solution space on a design prior to layout. Final simulations should always be done with fully coupled models.

Comparison to fully coupled simulations and measurements have shown that the accuracy of the SLEM model (for three lines) is adequate for the cross sections with a spacing/height ratio greater than 1. When this ratio is less than 1, the SLEM approximation should not be used. The method is always accurate, however, for two lines.

POINTS TO REMEMBER

- When estimating the effect of crosstalk in a system, the nearest neighbors have the greatest effect. The effects of the other lines fall off exponentially.
- The SLEM method should be used early in the design stage to quickly get a handle on the effect of crosstalk. SLEM is useful because it is computationally quick. Fully coupled simulations should always be performed on the final design.
- Common mode (all bits in phase) and differential mode (target bit out of phase) will produce the worst-case impedance and velocity variations.
- The accuracy of a three-line SLEM model falls off when the edge-to-edge spacing/height (above the ground plane) ratio is less than 1.

3.7. CROSSTALK TRENDS

As described in Section 3.6, the impedance variations due to crosstalk depend on the magnitude of the mutual capacitance and inductance. In turn, since the mutual parasitics will depend heavily on the cross-sectional geometry of the trace, the impedance variation will also. Lower-impedance traces will tend to exhibit less crosstalk-induced impedance variation than will high-impedance traces for a given dielectric constant. This is because lower-impedance traces exhibit heavier coupling to the reference planes (in the form of a higher capacitance, see Chapter 2). A transmission line that is strongly coupled to the reference planes will simply exhibit less coupling to adjacent traces. Lower-impedance lines are usually achieved either by widening the trace width or using a thinner dielectric. However, there are negative aspects to each of these choices. Wide traces will consume more routing area, and thin dielectrics can be very costly (thin dielectrics tend to have shorts between signal and reference plane layers, so the production yield is low).

Figure 3.13a and b show how even- and odd-mode impedance varies with trace-to-trace spacing. Notice that both the odd- and even-mode impedance values asymptotically approach the nominal impedance. The target line

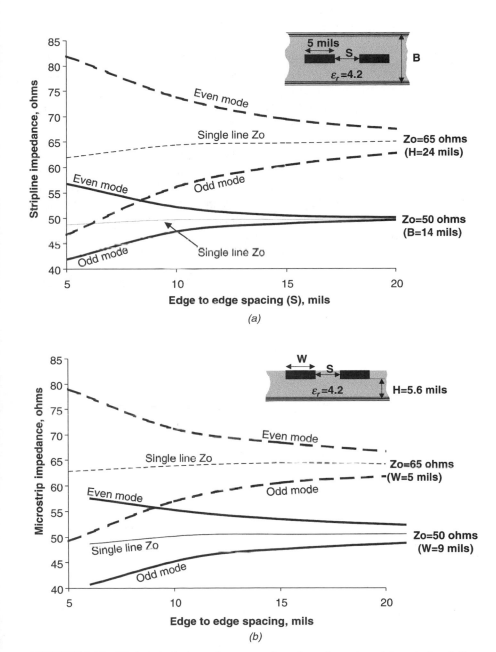

FIGURE 3.13 Variations in impedance as a function of spacing: (*a*) typical stripline two conductor system; (*b*) typical microstrip two-conductor system.

impedance was calculated with no consideration of the effect of adjacent traces (65 and 50 Ω). This is usually how impedance calculations are made and it is how board vendors test (they usually measure an isolated trace on a test coupon). Notice that at small spacing, the single-line impedance is lower than the target. This is because the adjacent traces increase the self-capacitance of the trace and effectively lower its impedance even when they are not active. When they are switching, the full odd- or even-mode impedance is realized. When designing high-trace-density boards, it is important to account for this effect.

The trace dimensions used to calculate the curves in Figure 3.14a and 3.14b were chosen to approximate real-world values for both 50- and 65-Ω trace values (both are common in the industry). Notice that the high-impedance traces exhibit significantly more impedance variation than do the lower-impedance traces because the reference planes are much farther in relation to the signal spacing. The mutual parasitics for these cross sections are plotted in Figure 3.14. Note that the mutual parasitics drop off exponentially with spacing. Also note that the values of the mutual terms are smaller for the lower impedance trace. The reader should note that these plots are for a simple two-conductor system. Significantly more variation can be expected when the target line is coupled to multiple aggressor nets.

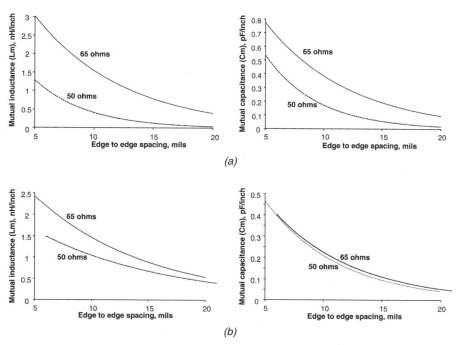

FIGURE 3.14 Mutual inductance and capacitance for (a) stripline in Figure 3.13a and (b) microstrip in Figure 3.13b.

POINTS TO REMEMBER

- Low-impedance line will produce less impedance variation from cross-talk for a given dielectric constant.
- The impedance of a single line on a board is influenced by the proximity of other traces even when they are not being actively driven.
- Mutual parasitics fall off exponentially with trace-to-trace spacing.

3.8. TERMINATION OF ODD- AND EVEN-MODE TRANSMISSION LINE PAIRS

The remaining issue that has not yet been covered is how to terminate a coupled transmission line pair. In most aspects of a design, it is usually prudent to minimize coupling between all traces. In some designs, however, this is either not possible or it is beneficial to have a high degree of coupling between two traces. One example where it is beneficial to have a high amount of coupling between traces is when the system clock on a computer is routed differentially. A differential transmission line consists of two tightly coupled traces that are always propagating in odd mode. The receiving circuitry is either a differential or a single-ended buffer. The input stage of a differential receiver usually consists of a differential pair that triggers at the crossing point of the two signals when the difference between the two signals is zero. Differential transmission lines are beneficial because they tend to exhibit better signal integrity than that of a single trace because they are more immune to noise and EMI is greatly reduced.

3.8.1. Pi Termination Network

One way to properly terminate a two-line coupled pair and prevent reflection in both the odd and even modes is to use a pi network. This particular termination scheme is useful when differential receivers are used. Refer to Figure 3.15. The resistances R_1, R_2, and R_3 must be chosen so that both the even and odd propagation modes will be terminated. First, let's consider even mode, where $V_1 = V_2 = V_e$. Since the voltage differences at nodes 1 and 2 will always be equal, no current will flow between these points. Subsequently, R_1 and R_2 must equal the even-mode impedance. To determine the value of R_3, it is necessary to evaluate the odd-mode propagation. Since V_1 and V_2 will always be equal but opposite values in the odd mode ($V_1 = -V_2 = V_O$), R_3 can be broken down into two series resistances, each with a value equal to, $\frac{1}{2}R_3$. For odd-mode propagation, the center connection of the two series resistors is a virtual ac ground. Refer to Figure 3.16 for a visual representation of the equivalent termination for the odd-mode case. To understand why the middle

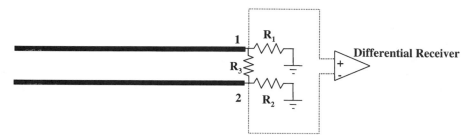

FIGURE 3.15   Pi termination configuration for a coupled transmission line pair.

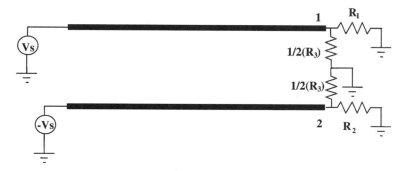

FIGURE 3.16   Equivalent of termination seen in the odd mode with the pi termination configuration.

of the resistor, R_3, must be ground for odd-mode propagation, consider a resistor that is constructed from a long piece of resistive material. If a voltage of 1.0 V is applied at one end and a voltage of -1.0 V is applied at the other end, the center of the resistor would have a voltage of 0.0 V. This means that each signal, in odd-mode propagation, is terminated with a value of R_1 (or R_2), in parallel with $\frac{1}{2}R_3$. The required values of R_1 and R_3 that may be used to terminate a pair of tightly coupled transmission lines in both even and odd propagation modes are

$$R_1 = R_2 = Z_{even} \tag{3.35}$$

$$R_3 = \frac{2(Z_{even})(Z_{odd})}{Z_{even} - Z_{odd}} \tag{3.36}$$

It should be noted, however, that if the transmission line pair will always be propagating in one mode (such as in the case of a differential clock line), the middle resistor, R_3, is not necessary.

3.8.2. T Termination Network

Another method of terminating both odd and even modes is a T-network of resistors. Refer to Figure 3.17. This particular termination scheme is useful

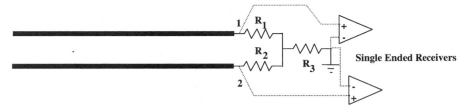

FIGURE 3.17   T termination configuration for a coupled transmission line pair.

FIGURE 3.18   Equivalent of termination seen in the even mode with the T termination configuration.

when a differential transmission line pair is used in conjunction with a single-ended receiver. First, let's consider the termination necessary for the odd mode. Because the voltage at node 1 will always be equal but opposite the voltage at node 2, we can place a virtual ac ground at the center, where R_3 is connected. This means that for odd-mode propagation, the lines are effectively terminated in the values of R_1 and R_2. Therefore, R_1 and R_2 must be set to a value equal to the odd-mode impedance. Now let's consider the even mode. Figure 3.18 shows how two resistors in parallel with a value of $2R_3$ are used to represent R_3. Since no current will flow between points 1 and 2, the resistor R_1 or R_2 in series with $2R_3$ must equal the even-mode impedance. The subsequent values for R_1, R_2, and R_3 in a T termination network are

$$R_1 = R_2 = Z_{\text{odd}} \tag{3.37}$$

$$R_3 = \tfrac{1}{2}(Z_{\text{even}} - Z_{\text{odd}}) \tag{3.38}$$

3.9. MINIMIZATION OF CROSSTALK

Since crosstalk is prevalent in virtually every aspect of a design, and its effects can be quite detrimental to system performance, it is useful to present some general guidelines to help reduce crosstalk. It should be noted, however, that all these guidelines will have adverse effects on the routability of the printed

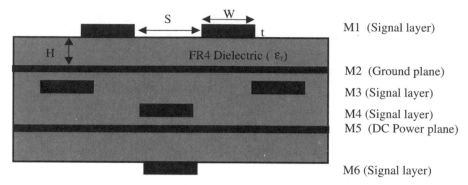

FIGURE 3.19   Dimensions that influence crosstalk.

circuit board. Since most circuit board designs are required to fit some pre-defined aspect ratio (i.e., a motherboard may be required to fit in a standard ATX chassis), a certain degree of crosstalk impact is inevitable. The following guidelines, however, will help the designer limit the negative impact of crosstalk. Refer to Figure 3.19 when reading the following list.

1. Widen the spacing S between the lines as much as routing restrictions will allow.
2. Design the transmission line so that the conductor is as close to the ground plane as possible (i.e., minimize H) while achieving the target impedance of the design. This will couple the transmission line tightly to the ground plane and less to adjacent signals.
3. Use differential routing techniques for critical nets, such as the system clock if system design allows.
4. If there is significant coupling between signals on different layers (such as layers M_3 and M_4), route them orthogonal to each other.
5. If possible, route the signals on a stripline layer or as an embedded microstrip to eliminate velocity variations.
6. Minimize parallel run lengths between signals. Route with short parallel sections and minimize long coupled sections between nets.
7. Place the components on the board to minimize congestion of traces.
8. Use slower edge rates. This, however, should be done with extreme caution. There are several negative consequences associated with using slow edge rates.

3.10. ADDITIONAL EXAMPLES

This example will build on the additional example in Chapter 2 and incorporate the effects of crosstalk into the high-speed bus example. Crosstalk often places severe limits on system performance, so it is important to understand the concepts presented in this chapter.

3.10.1. Problem

Assume that two components, U_1 and U_2, need to communicate with each other via an 8-bit-wide high-speed digital bus. The components are mounted on a standard four-layer motherboard with the stackup shown in Figure 3.20. The driving buffers on component U_1 have an impedance of 30 Ω and a swing of 0 to 2 V. The traces on the printed circuit board (PCB) are required to be 5 in. long with center-to-center spacing of 15 mils and impedance 50 Ω (ignoring crosstalk). The relative dielectric constant of the board (ε_r) is 4.0, the transmission line is assumed to be a perfect conductor, and the receiver capacitance is small enough to be ignored. Figure 3.21 depicts the circuit topology.

The transmission line parasitics are

mutual inductance = 0.54 nH/in.

mutual capacitance − 0.079 pF/in.

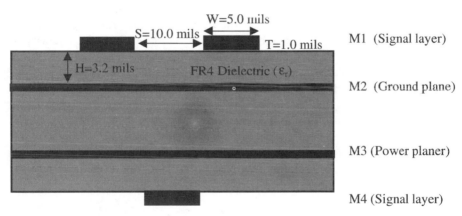

FIGURE 3.20 Cross section of PCB board used in the example.

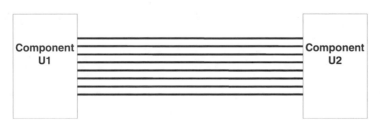

FIGURE 3.21 Circuit topology.

$$\text{self-inductance} = 7.13 \text{ nH/in.} \quad \text{(from the Chapter 2 example)}$$

$$\text{self-capacitance} = 2.85 \text{ pF/in.} \quad \text{(from the Chapter 2 example)}$$

3.10.2. Goals

1. Determine the maximum impedance variation on the transmission lines due to crosstalk.
2. Determine the maximum velocity difference due to crosstalk.
3. Assuming that the input buffers at component U_2 will switch at 1.0 V, determine if the buffer will false trigger due to crosstalk effects.

3.10.3. Determining the Maximum Crosstalk-Induced Impedance and Velocity Swing

As discussed in Section 3.6.2, the equivalent impedance of a transmission line in a coupled system is dependent on the switching pattern on the adjacent traces. To determine the worst-case impedance swing due to crosstalk, it is necessary to pick the line in the bus that will experience the most crosstalk. A line in the center of the bus is usually the best choice. As mentioned earlier in the chapter, only the nearest-neighboring lines will contribute the majority of the crosstalk effects. Therefore, the procedures illustrated in previous examples can be used to calculate the impedance variation due to crosstalk. The patterns that produce the worst-case crosstalk effects will always be either common mode or differential mode. In *common mode* the nets are all switching in phase, and in *differential mode* the target line is always switching 180° out of phase with the other nets on the bus.

First, let's evaluate the effects of common-mode propagation (all bits switching in phase). Assume that line 2 in Figure 3.22 represents a conductor in the middle of the 8-bit bus.

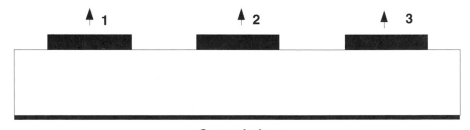

Ground plane

FIGURE 3.22 Common-mode switching pattern.

Even-mode capacitance of conductors 2 and $1 = C_{22} - C_{12}$
Even-mode capacitance of conductors 2 and $3 = C_{22} - C_{23}$
Equivalent capacitance of conductor $2 = C_{22} - C_{12} - C_{23}$
Even-mode inductance of conductors 2 and $1 = L_{22} + L_{12}$
Even-mode inductance of conductors 2 and $3 = L_{22} + L_{23}$
Equivalent inductance of conductor $2 = L_{22} + L_{12} + L_{23}$

Therefore,

$$Z_{2,\text{common}} = \sqrt{\frac{L_{22} + L_{12} + L_{23}}{C_{22} - C_{12} - C_{23}}}$$

$$= \sqrt{\frac{7.13\ \text{nH} + 0.54\ \text{nH} + 0.54\ \text{nH}}{2.85\ \text{pF} - 0.079\ \text{pF} - 0.079\ \text{pF}}} = 55.26\ \Omega$$

$\text{TD}_{2,\text{common}}$

$$= \sqrt{(L_{22} + L_{12} + L_{23})(C_{22} - C_{12} - C_{23})}$$

$$= \sqrt{(7.13\ \text{nH} + 0.54\ \text{nH} + 0.54\ \text{nH})(2.85\ \text{pF} - 0.079\ \text{pF} - 0.079\ \text{pF})}$$

$$= 148.6\ \text{pF/in.}$$

Now, let's evaluate the effects of differential-mode propagation. Assume that line 2 in Figure 3.23 represents a conductor in the middle of the 8-bit bus.

Odd-mode capacitance of conductors 2 and $1 = C_{22} + C_{12}$
Odd-mode capacitance of conductors 2 and $3 = C_{22} + C_{23}$
Equivalent capacitance of conductor $2 = C_{22} + C_{12} + C_{23}$
Even-mode inductance of conductors 2 and $1 = L_{22} - L_{12}$
Even-mode inductance of conductors 2 and $3 = L_{22} - L_{23}$
Equivalent inductance of conductor $2 = L_{22} - L_{12} - L_{23}$

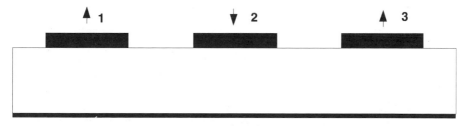

Ground plane

FIGURE 3.23 Differential switching pattern.

Therefore,

$$Z_{2,\text{differential}} = \sqrt{\frac{L_{22} - L_{12} - L_{23}}{C_{22} + C_{12} + C_{23}}}$$

$$= \sqrt{\frac{7.13 \text{ nH} - 0.54 \text{ nH} - 0.54 \text{ nH}}{2.85 \text{ pF} + 0.079 \text{ pF} + 0.079 \text{ pF}}} = 44.8 \ \Omega$$

$\text{TD}_{2,\text{differential}}$

$$= \sqrt{(L_{22} - L_{12} - L_{23})(C_{22} + C_{12} + C_{23})}$$

$$= \sqrt{(7.13 \text{ nH} - 0.54 \text{ nH} - 0.54 \text{ nH})(2.85 \text{ pF} + 0.079 \text{ pF} + 0.079 \text{ pF})}$$

$$= 135 \text{ ps/in.}$$

The velocity and impedance variations due to crosstalk are as follows:

$$44.8 \ \Omega < Z_o < 55.26 \ \Omega$$

$$135 \text{ ps/in.} < \text{TD} < 148.6 \text{ ps/in.}$$

3.10.4. Determining if Crosstalk Will Induce False Triggers

A false trigger will occur if the signal has severe signal integrity problems. A receiver will trigger at its threshold. Typical CMOS buffers have a threshold voltage at $\frac{1}{2}V_{cc}$. If the signal rings back below the threshold level at the receiver, the signal integrity may cause a false trigger. If this happens in a digital system, it could cause a wide variety of problems, ranging from glitches in the data to catastrophic system failure (especially if the false trigger is on a strobe or clock signal).

To determine if the crosstalk will cause a false signal, it is necessary to evaluate the signal integrity. There are several methods of doing this. The simplest method is to use a simple bounce (i.e., lattice) diagram, such as was used to determine the wave shape of the signal in the Chapter 2 example. The full bounce diagram, however, is not required. Since the second reflection seen at the receiver will produce the highest magnitude of ringing, all that is required is the voltage level cause by the second reflection as seen at the receiver. Furthermore, the most ringing will occur when the driver impedance is low and the transmission line impedance is high (see Chapter 2). Therefore, the worst-case ringing will be achieved when the coupled lines in the bus are propagating in phase with each other (i.e., common mode). The common-mode impedance was calculated above to be 55.2 Ω. The signal integrity is calculated using a bounce diagram as shown in Figure 3.24.

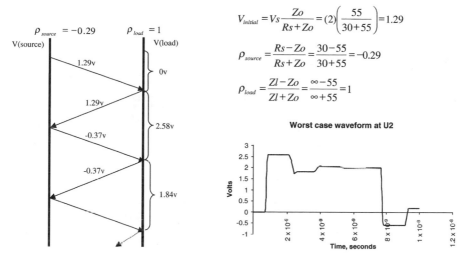

$$V_{initial} = Vs\frac{Zo}{Rs+Zo} = (2)\left(\frac{55}{30+55}\right) = 1.29$$

$$\rho_{source} = \frac{Rs-Zo}{Rs+Zo} = \frac{30-55}{30+55} = -0.29$$

$$\rho_{load} = \frac{Zl-Zo}{Zl+Zo} = \frac{\infty-55}{\infty+55} = 1$$

FIGURE 3.24 Calculation of the final waveform.

The signal will ring down to a minimum voltage of 1.84 V. Since the threshold voltage of this buffer was defined to be 1.0 V, and the signal does not ring back down below this value, the signal integrity will not cause a false triggering of the receiver.

CHAPTER 4

Nonideal Interconnect Issues

Modern technology has shown a relentless trend toward higher speeds and smaller form factors. Subsequently, effects previously considered to be negligible and ignored during digital design often become primary design issues. Nonideal effects such as frequency-dependent losses, impedance discontinuities, and serpentine effects are just a few of the new variables that must be accounted for in modern designs. Some of these high-frequency effects are very difficult to model and are continuously being researched by universities. Subsequently, as system designs progress in speed, the engineer not only has to deal with technically difficult issues but must also contend with a significantly greater number of variables. In this chapter we address several of the dominant nonideal interconnect issues that must be addressed in modern designs. The focus of this chapter is on the high-speed transmission characteristics that have been largely ignored in designs of the past but are critical in modern times. Many of the models presented here have known shortcomings that will, much like the simpler models of previous days, need to be revised sometime in the future. As with any set of models, the user should continually be aware of the approximations and assumptions being made.

4.1. TRANSMISSION LINE LOSSES

As digital systems evolve and technology pushes for smaller and faster systems, the geometric dimensions of the transmission lines and package components are shrinking. Smaller dimensions and high-frequency content cause the resistive losses in the transmission line to be exacerbated. Modeling the resistive losses in transmission lines is becoming increasingly important. Resistive losses will affect the performance of a digital system by decreasing the signal amplitude, thus affecting noise margins and slowing edge rates, which in turn affects timing margins. Previously it has been possible to ignore losses on the PCB and in the package because systems operated at slower frequencies. Modern systems, however, require rigorous analysis of losses because they are often a first-order effect that significantly degrades the performance of digital interconnects.

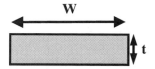

FIGURE 4.1   Microstrip line current density at dc. At dc, current flows through entire area of the cross section where area = $A = Wt$.

4.1.1. Conductor DC Losses

As mentioned in Chapter 2, there is a resistive component in the transmission line model. This resistive component exists because the conductors used to manufacture the transmission lines on a PCB are not perfect conductors. The loss in microstrip and stripline conductors can be broken down into two components: dc and ac losses. Dc losses are of particular concern in small-geometry conductors, very long lines, and multiload (also known as multidrop) buses. Long copper telecommunication lines, for example, must have repeaters every few miles to receive and retransmit data because of signal degradation. Additionally, designs of multiprocessor computer systems experience resistive drops that can encroach on logic threshold levels and reduce noise margins.

The dc loss depends primarily on two factors: the resistivity of the conductor and the total area in which the current is flowing. Figure 4.1 shows the current distribution on a microstrip line at dc (0 Hz). The current flows through the entire cross section of the conductor, and the resistive loss can be found using the equation

$$R = \frac{\rho L}{A} = \frac{\rho L}{Wt} \qquad \text{ohms} \qquad (4.1)$$

where R is the total resistance of the line, ρ the resistivity of the conductor material in ohm-meters (the inverse of conductivity), L the length of the line, W the conductor width, t the conductor thickness, and A the cross-sectional area of the signal conductor. The losses in the ground return path in a conventional design are usually negligible at dc because the cross-sectional area is very large compared to the signal line.

4.1.2. Dielectric DC Losses

Since the dielectric materials used in PCBs are not perfect insulators, there is a dc loss associated with the resistive drop across the dielectric material between the signal conductor and the reference plane. The dielectric losses at dc for conventional substrates, however, are usually very negligible and can be ignored. Frequency-dependent dielectric losses are discussed in Section 4.1.4.

4.1.3. Skin Effect

At low frequencies it is adequate to use only dc losses in system simulation, however, as the frequency increases, other phenomena that vary with the spectral content of the digital signals begin to dominate. The most prominent of these frequency-dependent variables is the *skin effect*, so named because at high frequencies, the current flowing in a conductor will migrate toward the periphery or "skin" of the conductor.

Frequency-Dependent Resistance and Inductance. Skin effect manifests itself primarily as resistance and inductance variations. At low frequencies, the resistance and inductance assume dc values, but as frequency increases, the cross-sectional current distribution in the transmission line becomes nonuniform and moves to the exterior of the conductor. The changing current distribution causes the resistance to increase with the square root of frequency and the total inductance to fall asymptotically toward a static value called the *external inductance*.

To understand how this happens, imagine a signal traveling on a microstrip transmission line. The cross section in Figure 2.3 shows that a high-frequency signal travels between the signal trace and the reference plane in the form of time-varying electric and magnetic fields and is not contained wholly inside the conductor. When these fields intersect the signal trace or the ground plane conductor, they will penetrate the metal and their amplitudes will be attenuated. The amount of attenuation will depend on the resitivity (or conductivity) of the metal and the frequency content of the signal. The amount of penetration into the metal, known as the *skin depth*, is usually represented by the symbol δ. If the metal is a perfect conductor, there will be no penetration into the metal and the skin depth will be zero. If the metal has a finite conductivity, the portion of the electromagnetic wave that penetrates the metal will be attenuated so that at the skin depth, its amplitude will be decayed by a factor of e^{-1} of its initial value at the surface. (If the reader is interested in the derivation of the skin depth, any standard electromagnetic textbook should suffice.) This will confine the signal so that approximately 63% of the total current will flow in one skin depth and the current density will decay exponentially into the thickness of the conductor. The physical dimensions of the conductor, of course, modify this current distribution. The skin depth δ is calculated with the equation

$$\delta = \sqrt{\frac{2\rho}{\omega\mu}} = \sqrt{\frac{\rho}{\pi F \mu}} \qquad \text{meters} \qquad (4.2)$$

where ρ is the resistivity (the inverse of the conductivity) of the metal, ω the angular frequency ($2\pi F$), and μ the permeability of free space (in henries per meter) [Johnk, 1988]. (In the rare circumstance that a magnetic metal such as iron or nickel is used, the appropriate permeability should be substituted. This will not occur except in very unusual situations.) It can be seen that as the frequency increases, the current will be confined to a smaller area, which will increase the resistance, as seen in the dc resistance calculation (4.1).

This phenomenon also causes variations of the inductance with the frequency. The total inductance of a conductor is caused by magnetic flux produced from the current. Some of the flux is produced from the current in the wire itself. Since at high frequencies the current will be confined primarily to the skin depth, the current will cease to propagate in the center of the conductor. Subsequently, the internal inductance (so named for that portion of current flowing in the internal portion of the conductor) will become negligible and the total inductance will decrease. External inductance is the value calculated when it is assumed that all the current is flowing on the exterior of the conductor. The total inductance is obtained by summing the external and internal inductances. In most high-speed digital systems, the frequency components of the signals are high enough so that it is a valid approximation to ignore the frequency dependence of inductance. Therefore, for the remainder of this book, only the external inductance is considered.

Frequency-Dependent Conductor Losses in a Microstrip.

By extending the dc resistance equation (4.1), the frequency dependence of the resistance in a transmission line conductor can be approximated. Frequency-dependent resistance will sometimes be referred to ac resistance in this book. At low frequencies, the ac resistance will be identical to the dc resistance because the skin depth will be much greater than the thickness of the conductor. The ac resistance will remain approximately equal to the dc resistance until the frequency increases to a point where the skin depth is smaller than the conductor thickness. Figure 4.2 depicts the current distribution on a microstrip line at high frequencies. Notice that the current distribution is concentrated on the bottom edge of the transmission line. This is because the fields between the signal line and the ground plane pull the charge to the bottom edge. Also notice that the current distribution curves up the side of the conductor. This is because there is still significant field concentration along the thickness (the t dimension in Figure 4.2) of the conductor. The amount of cross-sectional area

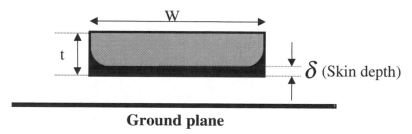

Ground plane

FIGURE 4.2 Current distribution on a microstrip transmission line. 63% of the current is concentrated in the darkly shaded area due to the skin effect.

in which the current is flowing will become smaller as the frequency increases [see equation (4.2)].

The losses in the conductor can be approximated using the dc resistance and the skin effect formulas by substituting the skin depth for the conductor thickness:

$$R_{ac\ signal} \approx \frac{\rho}{W\delta} = \frac{\sqrt{\rho\pi\mu F}}{W} \qquad \Omega/m \qquad\qquad (4.3a)$$

Note that the approximation is valid only when the skin depth is smaller than the conductor thickness. Furthermore, equation (4.3a) is only an approximation because it assumes that all the current is flowing in the skin depth and is presented in this form for instructional purposes only. As this section progresses, more accurate methods of calculating the ac losses will be presented. Notice that when the skin depth equation (4.2) is inserted into the dc resistance equation, the ac resistance becomes directly proportional to the square root of the frequency F and the resistivity ρ. Note that in equation (4.3a) the length term has been excluded to give the ac resistance units of resistance per unit length.

Figure 4.3 is a plot of the skin depth versus frequency for a copper conductor. Note that the skin depth is greater than the conductor thickness at frequencies below approximately 1.7 MHz. Figure 4.4 is a plot of resistance as a function of frequency for the example copper cross section. Note that the initial portion of the curve is constant at the dc resistance. This section

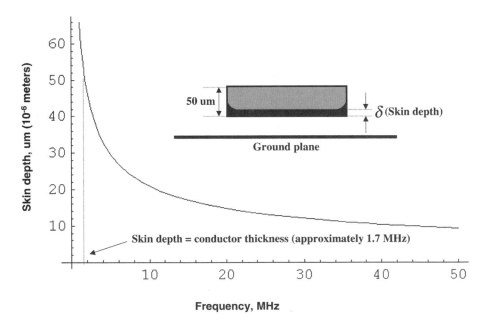

FIGURE 4.3   Skin depth as a function of frequency.

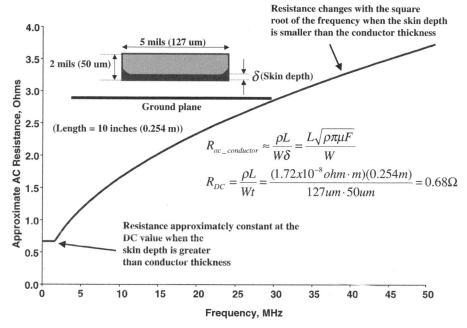

FIGURE 4.4 Ac resistance as a function of frequency.

corresponds to frequencies where the skin depth is greater than the conductor thickness. The curve begins to change with the square root of frequency when the skin depth becomes smaller than the conductor thickness. Although the curve shown in Figure 4.4 is not based on an exact model, it is well suited to help the reader understand the fundamental behavior of skin effect resistance. A good way to match measurements when simulating both the ac and dc resistance of a transmission line in a simulator such as SPICE is to combine R_{ac} and R_{dc}:

$$R_{\text{total}} \approx \sqrt{R_{ac}^2 + R_{dc}^2}$$ (4.3b)

The skin effect resistance of the conductor, however, is only one part of the total ac resistance. The portion that is not included in equation (4.3a) is the resistance of the return current on the reference plane. The return current will flow underneath the signal line in the reference plane and will be concentrated largely in one skin depth and will spread out perpendicular to the trace direction, with the highest amount of current concentrated directly beneath the signal conductor. An approximate current density distribution in the ground plane for a microstrip transmission line is [Johnson and Graham, 1993]

$$I(D) \approx \frac{I_o}{\pi H} \frac{1}{1 + (D/H)^2}$$ (4.4)

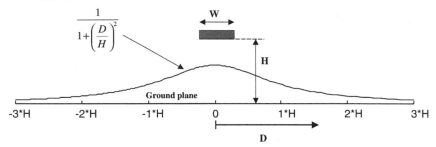

FIGURE 4.5 Current density distribution in the ground plane.

where I_o is the total signal current, D the distance from the trace (see Figure 4.5), and H the height above the ground plane. Figure 4.5 is a graphical representation of this current density distribution.

An approximation of the effective resistance of the ground plane can be derived using a technique similar to that used to find the ac resistance of the signal conductor. First, since 63% of the current will be confined to one skin depth (δ), then for the resistance calculation, the approximation may be made that the ground current flows entirely in one skin depth, as was approximated for the signal conductor ac resistance. Second, the equation

$$\int_{-3H}^{3H} \frac{I_o}{\pi H} \frac{1}{1 + (D/H)^2} = \frac{2I_o}{\pi} \tan^{-1}(3) \approx 0.795 I_o$$

$$= 79.5\% \text{ of the total current} \qquad (4.5)$$

shows that 79.5% of the current is contained within a distance of $\pm 3H$ ($6H$ total width) away from the center of the conductor. Thus, the ground return path resistance can be approximated by a conductor of cross section $A_{\text{ground}} = \delta \times 6H$. Substituting this result into equation (4.1) yields

$$R_{\text{ac ground}} \approx \frac{\rho}{A_{\text{ground}}} = \frac{\rho}{6\delta H} = \frac{\rho}{6H} \sqrt{\frac{\pi \mu F}{\rho}}$$

$$= \frac{\sqrt{\rho \pi \mu F}}{6H} \qquad \Omega/\text{m} \qquad (4.6)$$

The total ac resistance is the sum of the conductor and ground plane resistance:

$$R_{\text{ac microstrip}} = R_{\text{ac signal}} + R_{\text{ac ground}} \qquad (4.7)$$

$$\approx \frac{\sqrt{\rho \pi \mu F}}{W} + \frac{\sqrt{\rho \pi \mu F}}{6H} = \sqrt{\rho \pi \mu F} \left(\frac{1}{W} + \frac{1}{6H} \right) \qquad \Omega/\text{m}$$

$$(4.8)$$

Equation (4.8) should be considered a first-order approximation. However, since surface roughness can increase resistance by 10 to 50% (see "Effect of Conductor Surface Roughness" below), equation (4.8) will probably provide an adequate level of accuracy for most situations.

A more exact formula for the ac resistance of a microstrip can be derived through conformal mapping techniques.

$$R_{signal} = [\text{loss ratio}] \left(\frac{1}{\pi} + \frac{1}{\pi^2} \ln \frac{4\pi W}{t} \right) \frac{\sqrt{\pi \mu F \rho}}{W}$$

$$\text{Loss ratio} = \begin{cases} 0.94 + 0.132\dfrac{W}{H} - 0.0062 \left(\dfrac{W}{H} \right)^2 & \text{for} \quad 0.5 < \dfrac{W}{H} < 10 \\[2ex] 1 & \text{for} \quad 0.5 \geq \dfrac{W}{H} \quad (4.9) \end{cases}$$

$$R_{ground} = \frac{W/H}{W/H + 5.8 + 0.03(H/W)} \frac{\sqrt{\pi \mu F \rho}}{W} \quad \text{for} \quad 0.1 < \frac{W}{H} < 10$$

$$R_{ac\ microstrip} = R_{signal} + R_{ground} \quad \Omega/\text{m}$$

Equation set (4.9) was derived using conformal mapping techniques and appears to have excellent agreement with experimental results [Collins, 1992]. These formulas are significantly more cumbersome than (4.8) but should yield the most accurate results. Equation (4.8) will tend to yield resistance values that are larger then those in (4.9). Often, the slightly larger values given by (4.8) are used to roughly approximate the additional resistance gained from surface roughness.

Frequency-Dependent Conductor Losses in a Stripline.

In a stripline transmission line, the currents of a high-frequency signal are concentrated in the upper and lower edges of the conductor. The current density will depend on the proximity of the local ground planes. If the stripline is referenced to two ground planes equidistant from the conductor, for example, the current will be divided equally in the upper and lower portions of the conductor, as depicted in Figure 4.6. In an offset transmission line, the current densities on the upper and lower edges of the transmission line will depend on the relative distances between the ground planes and the conductor (H_1 and H_2 in Figure 4.6). The current density distributions in each of the return planes in a stripline will be governed by an equation similar to equation (4.4) and will differ only in the magnitude of the term I_o, which will obviously be a function of the respective distances of the reference planes. Thus, the resistance of a stripline can be approximated by the parallel combination of the resistance in the top and bottom portions of the conductor. The resistance equations for the upper and lower sections of the stripline may be obtained by applying equation (4.8)

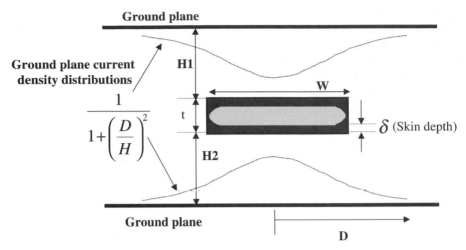

FIGURE 4.6 Current density distribution in a stripline.

or (4.9) for the appropriate value of H. These two resistance values must then be taken in parallel to obtain the total resistance for a stripline [see equation (4.10)].

A good approximation of the ac resistance of a stripline is

$$R_{\text{ac stripline}} = \frac{(R_{(H1)\text{ac microstrip}})(R_{(H2)\text{ac microstrip}})}{R_{(H1)\text{ac microstrip}} + R_{(H2)\text{ac microstrip}}} \quad \Omega/\text{m} \qquad (4.10)$$

where the microstrip resistance values are from equation (4.8) or (4.9) evaluated at heights H_1 or H_2. Refer to Figure 4.6 for the relevant dimensions.

Effect of Conductor Surface Roughness. As discussed earlier, high-frequency signals will experience increased series resistance at high frequency due to migration of the current toward the conductor surface. The formulas for these losses, however, are derived on the assumption of perfectly smooth metal surfaces. In reality, the metal surfaces will be rough, which will effectively increase the resistance of the material when the mean surface roughness is a significant percentage of the skin depth. In experimental studies it has been found that high-frequency signals traveling on lines with significant surface roughness exhibit losses that are higher than those calculated with the ideal formulas by as much as 10 to 50%. Because the roughness pattern is random, it is impossible to predict the skin effect losses exactly. However, by observing the magnitude of the roughness compared to the skin depth, it is easy to determine if the surface roughness will be of a dominant consequence. Figure 4.7 is a drawing of what a stripline in a typical PCB would look like if it were cross-sectioned and examined with a microscope. Note

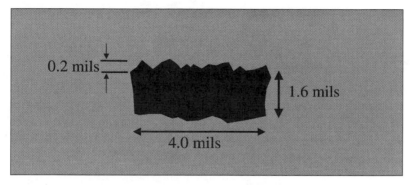

FIGURE 4.7   Cross section of a stripline in a typical PCB, showing surface roughness.

that in this particular case, the surface roughness on the top of the conductor is approximately 0.2 mil (5 μm).

The roughness of the conductor is usually described as the *tooth structure* and the magnitude of the surface variations is described as *tooth size*. For example, the measured tooth size in Figure 4.7 is 5 μm. A poll of PCB vendors indicate that typical FR4 boards will have an average tooth size of 4 to 7 μm. Conductor surfaces that are exposed to the etching process will usually have a significantly smaller magnitude of tooth size. Subsequently, the tooth structure with the largest tooth size is usually located on the side of the conductor that is facing the reference plane. Notice that the lower side of the conductor depicted in Figure 4.7 has a significantly smaller tooth size, which corresponds to the etched side of the conductor. There are processes that will significantly decrease the amount of surface roughness facing the reference plane; however, they tend to be expensive for high-volume manufacturing purposes at the time of this writing.

The surface roughness will begin to affect the accuracy of the ideal ac resistance equations when the magnitude of the tooth size becomes significant compared to the skin depth. For example, when the frequency reaches 200 MHz, the skin depth in copper will be approximately equal to the surface roughness of a typical PCB. Spectral components above this frequency will experience increasing deviation from the ideal formulas. To gauge the magnitude of the surface roughness, it is necessary either to perform a cross-sectional measurement such as the one presented in Figure 4.7, or to ask the PCB manufacturer for the roughness specification. If the surface roughness is a concern, the ac resistance must be measured.

Frequency-Dependent Properties of Various Metals.   The surface resistance (R_s) is often the parameter used to describe the ac resistance of a given material. The surface resistance is simply the ac resistance as calculated with equations (4.3) through (4.9) with the square root of the frequency divided

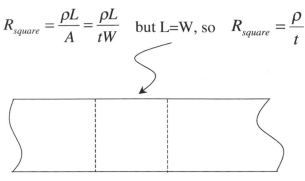

$$R_{square} = \frac{\rho L}{A} = \frac{\rho L}{tW} \quad \text{but L=W, so} \quad R_{square} = \frac{\rho}{t}$$

FIGURE 4.8   Concept of ohms/square.

out. Subsequently, the ac resistance will take the form

$$R_{ac} = R_s \sqrt{F} \tag{4.11}$$

For the purpose of material property classification, the surface resistance is often calculated for a semi-infinite plane with the length equal to the width ($L/W = 1$). It is also assumed that the resistance is calculated in the same manner as in equation (4.3), where all the current is confined to one skin depth. This approximation is shown in equation (4.12b).

$$R_s = \begin{cases} \dfrac{\rho}{\delta} = \dfrac{L}{W}\sqrt{\rho\pi\mu} & \text{ohms} \cdot \sqrt{\text{seconds}} & (4.12a) \\[2mm] \sqrt{\rho\pi\mu} & \text{ohms} \cdot \sqrt{s}/\text{square} \quad (L=W) & (4.12b) \end{cases}$$

For a finite area of conductor, the ac resistance is obtained by multiplying R_s (in ohms per square) by the length and the square root of the frequency and dividing by the width. The units of R_{ac} are often reported in ohms per square. Note that the "per square" terminology refers simply to a square of any size (i.e., length = width), since the resistance of one geometric square of board trace at a given thickness is independent of the size of the square. Thus, the surface resistance can be estimated simply by counting how many squares can be fit geometrically into the area under examination and multiplying that number by R_s times the square root of frequency. The concept of dc resistance per square is illustrated in Figure 4.8. Remember, in interconnect simulation it is very important to account for both the conductor and the ground return path resistance. Table 4.1 shows the skin effect properties of typical materials.

Effect of AC Losses on Signals. There are two classic implementations of ac losses: those for digital designers and those for microwave designers. The microwave designer is usually interested only in the ac resistance in a frequency-domain simulation. This is easy to implement because most general simulators, such as HSPICE, have a frequency-dependent resistor, which

TABLE 4.1. Frequency-Dependent Properties of Typical Metals

Metal	Resistivity ($\Omega \cdot m$)	Skin Depth (m)	Surface Resistivity, R_s ($\Omega \cdot \sqrt{s}$/square)
Copper (300 K)	1.72×10^{-8}	$0.066F^{-1/2}$	2.61×10^{-7}
Copper (77 K)	5.55×10^{-9}	$0.037F^{-1/2}$	1.5×10^{-7}
Brass	6.36×10^{-8}	$0.127F^{-1/2}$	5.01×10^{-7}
Silver	1.62×10^{-8}	$0.0642F^{-1/2}$	2.52×10^{-7}
Aluminum	2.68×10^{-8}	$0.0826F^{-1/2}$	3.26×10^{-7}

Source: Data from Ramo et al. [1994].

can be made to vary with the root of the frequency [see equation (4.11)] and used in an equivalent circuit constructed of cascaded *LRC* segments as in Figure 2.4.

The digital engineer, however, has a more difficult problem. Digital signals approximate square waves and are subsequently wide band, which means that they contain many frequency components. This is an important concept to understand. We demonstrate this with the equation [Selby, 1973]

$$f(x) = \frac{2}{\pi} \sum_{n=1,3,5,\ldots} \frac{1}{n} \sin 2\pi nFx \qquad (4.13)$$

which is the Fourier expansion of a periodic square wave at a 50% duty cycle, where F is the frequency and x is the time. A 100-MHz periodic square wave, for example, will be a superposition of an infinite number of sine waves of frequencies that are an odd multiple of the fundamental (i.e., 100 MHz, 300 MHz, 500 MHz, etc.). These components are referred to as *harmonics*, where $n = 1$ corresponds to the first harmonic, $n = 3$ corresponds to the third harmonic, and so on. The skin effect will cause each one of these harmonics to be attenuated with increasing frequency. A real-life signal will, of course, have additional frequency content due to the fact that it is not a perfect square wave with 50% duty cycle and infinitely fast rise times, as was assumed in equation (4.13). Notably, as the reader can verify with Fourier techniques, the even harmonics become present in the spectrum if the waveform is not 50% duty cycle. This will be come significant in Chapter 10.

Since the ac resistance is defined in terms of the frequency domain and digital signals are defined in terms of the time domain, it is difficult to account correctly for frequency-dependent losses in time-domain simulations. Fortunately, many simulators do a good job at this approximation assuming that the user inserts the correct value of R_s. The ac resistance is usually measured with a vector network analyzer (VNA). In the frequency domain, the ac resistance is often characterized by the attenuation factor, α, which is a measure of signal amplitude loss as a function of frequency across a

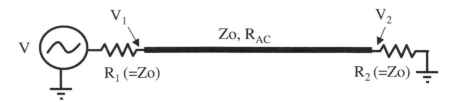

FIGURE 4.9   Schematic for equations (4.14a) to (4.14c).

transmission line. In a matched system, such as that depicted in Figure 4.9, the attenuation factor can easily be calculated at a single frequency as shown in equations (4.14a–c), which is based on the voltage divider at the receiver with the ac resistance included in the denominator. If the termination and driver impedance do not match the characteristic impedance of the transmission line, the voltage-divider method of equation (4.14c) is not valid because the reflections interfere with the measurement. Since the characteristic impedance of the transmission line does not usually exactly match the internal resistance of the VNA, techniques to eliminate the effect of the reflections must be used. Techniques for measurement of ac losses are described in Chapter 11.

$$V_1 = V \frac{R_{ac} + R_2}{R_1 + R_{ac} + R_2} \tag{4.14a}$$

$$V_2 = V \frac{R_2}{R_1 + R_2 + R_{ac}} \tag{4.14b}$$

$$\alpha = \frac{V_2}{V_1} = \frac{R_2}{R_{ac} + R_2}. \tag{4.14c}$$

To demonstrate the effect of frequency-dependent losses on a time-domain signal, refer to Figure 4.10. Figure 4.10a is an ideal trapezoidal wave that represents a digital signal. Figure 4.10b is the Fourier transform of the digital signal, showing the frequency spectrum contained in the signal. Figure 4.10c is a plot of attenuation versus frequency for a 10-in.-long microstrip transmission line. Figure 4.10d shows the waveform obtained when the attenuation curve is multiplied by the FFT (fast Fourier transform) curve and inverse FFT is performed. Notice that the wave shape is rounded, the edge rate has decayed, and the amplitude has been attenuated. These effects occur because the high-frequency components of the signal have been attenuated significantly, which eliminates the sharp edges and degrades the edge.

Each simulator will have its own way to implement the ac resistance. All will require the user to input the surface resistance (R_s) and either a cross section of the transmission line or the inductance and capacitance matrix (see Chapter 2).

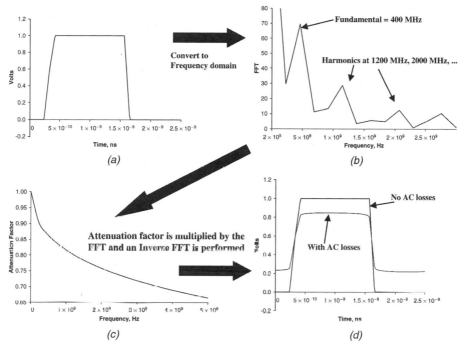

FIGURE 4.10 Effect of frequency-dependent losses on a time-domain signal: (*a*) ideal digital pulse (400 MHz periodic, pulse width = 1.25 ns, period = 2.5 ns); (*b*) Fourier transform showing frequency components; (*c*) attenuation factor versus frequency; (*d*) effect that frequency-dependent losses have a time-domain signal.

4.1.4. Frequency-Dependent Dielectric Losses

In most designs the dielectric losses can be ignored because the conductor losses are dominant. As frequencies increase, however, this assumption will cease to be true. Subsequently, it is important to understand the fundamental mechanisms that cause the dielectric losses to vary with frequency.

When a time-varying electric field is impressed onto a material, any molecules in the material that are polar in nature will tend to align in the direction opposite that of the applied field. This is called *electric polarization*. The classical model for dielectric losses, which was inspired by experimental measurement, involves an oscillating system of molecular particles, in which the response to the applied electric fields involves damping mechanisms that change with frequency [Johnk, 1988]. Any frequency variations in the dielectric loss are caused by these mechanisms.

When dielectric losses are accounted for, the dielectric constant of the material becomes complex:

$$\varepsilon = \varepsilon' - j\varepsilon'' \tag{4.15}$$

where the imaginary portion represents the losses and the real portion is the typical value of the dielectric constant. Since the imaginary portion of equation (4.15) represents the losses, it is convenient to think of it as the effective conductivity (i.e., the inverse of the resistivity) of the lossy dielectric.

Subsequently, as shown in any electromagnetics textbook, $1/\rho = 2\pi F\varepsilon''$ becomes the equivalent loss mechanism, where ρ is the effective resistivity of the dielectric material and F is the frequency. Unlike metals, however, the losses of a dielectric are usually not characterized by its resistivity. The typical method of loss characterization in dielectrics is by the loss tangent [Johnk, 1988]:

$$\tan|\delta_d| = \frac{1}{2\rho\pi F\varepsilon} = \frac{\varepsilon''}{\varepsilon'} \tag{4.16}$$

where ρ is the resistivity of the dielectric. For most practical applications, simulators allow the input of a loss tangent as in equation (4.16). However, if the designer desires to create an equivalent circuit of a transmission line with LRC segments as in Section 2.3.3, a relationship between $\tan|\delta_d|$ and the shunt resistance G in the transmission line model must be established. This relationship is shown as [Collins, 1992]

$$G = \frac{\varepsilon''}{\varepsilon'}(2\pi F C_{11}) \qquad \text{siemens} \tag{4.17}$$

where C_{11} is the self-capacitance per unit length and F is the frequency. The loss tangent will change with frequency and material properties. Figure 4.11 shows the loss tangent for a typical sample of FR4 as a function of frequency.

Example 4.1: Calculating Losses.   Refer to the stripline cross section depicted in Figure 4.6.

1. Calculate the surface resistivity (R_s) assuming that $W = 5$ mils, $H_1 = H_2 = 10$ mils, $t = 0.63$ mil, and $\varepsilon_r = 4.0$.
2. Determine how the resistance will change as a function of the frequency.
3. Calculate the shunt resistance due to the dielectric losses at 400 MHz.
4. Calculate the series resistance due to the conductor losses at 400 MHz.

SOLUTION: Equations (4.9), (4.10), and (4.11) are used to calculate the surface resistivity for the transmission line. Initially, the resistance for both the bottom and top portions of the transmission line is calculated assuming the microstrip equations. Then equation (4.10) is used to determine the stripline resistance. To determine the resistivity, the square root of the frequency is divided out.

$$\frac{W}{H} = \frac{5}{10} = 0.5$$

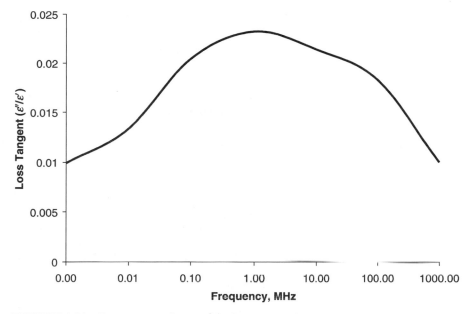

FIGURE 4.11 Frequency variance of the loss tangent in typical FR4 dielectric. (Adapted from Mumby [1988].)

Loss ratio = 1.0

$$R_{signal} = 1.0 \left(\frac{1}{\pi} + \frac{1}{\pi^2} \ln \frac{4\pi(5)}{0.63} \right) \frac{\sqrt{(\pi \cdot 12.56 \times 10^{-7})1.72 \times 10^{-8}(F)}}{(0.005 \text{ in.})(0.0254 \text{ m/in.})}$$

$$= 0.00172\sqrt{F} \qquad \Omega/\text{m}$$

$$R_{ground} = \left(\frac{1}{0.5 + 5.8 + 0.03 \cdot 0.5} \right) \frac{\sqrt{(\pi \cdot 12.56 \times 10^{-7})1.72 \times 10^{-8}(F)}}{(0.005 \text{ in.})(0.0254 \text{ m/in.})}$$

$$= 0.0003\sqrt{F} \qquad \Omega/\text{m}$$

$$R_{(H1)\text{ac microstrip}} = R_{signal} + R_{ground} = (0.00172 + 0.0003)\sqrt{F}$$

$$= 0.0020\sqrt{F} \qquad \Omega/\text{m}$$

$$R_{(H2)\text{ac microstrip}} = R_{(H1)\text{ac microstrip}}$$

$$R_{\text{ac stripline}} = \frac{(R_{(H1)\text{ac microstrip}})(R_{(H2)\text{ac microstrip}})}{R_{(H1)\text{ac microstrip}} + R_{(H2)\text{ac microstrip}}}$$

$$= \frac{R_{(H1)\text{ac microstrip}}}{2} = 0.0010\sqrt{F} \qquad \Omega/\text{m}$$

$$R_s = 0.0010 \frac{\Omega}{\text{m} \cdot \sqrt{\text{Hz}}}$$

To determine how the resistance will vary with frequency, it is necessary to determine the frequency at which the skin depth becomes smaller than the conductor thickness. To do so, we examine equation (4.2), substitute the conductor thickness in place of the skin depth, and solve for the frequency.

$$0.63 \text{ mil} \frac{25.4 \times 10^{-6} \text{ m}}{\text{mil}} = \sqrt{\frac{1.72 \times 10^{-8} \ \Omega \cdot \text{m}}{\pi [12.56 \times 10^{-7} \ (\text{H/m})]F}} \Rightarrow F = 17 \text{ MHz}$$

Below 17 MHz, the resistance of this conductor is approximately equal to the dc resistance [equation (4.1)]:

$$R_{\text{dc}} = \frac{1.72 \times 10^{-8} \ \Omega \cdot \text{m}}{5 \text{ mil} \dfrac{25.4 \times 10^{-6} \text{ m}}{\text{mil}} \left(0.63 \dfrac{25.4 \times 10^{-6} \text{ m}}{\text{mil}} \right)} = 8.5 \ \Omega/\text{m}$$

Above 17 MHz, the resistance will vary with the square root of frequency:

$$R_{\text{ac}} = (0.0010)\sqrt{F} \qquad \Omega/\text{m}$$

Therefore, the resistance at 400 MHz is

$$R_{\text{ac}}(400 \text{ MHz}) = 0.0010\sqrt{400 \times 10^6} = 20.2 \ \Omega/\text{m}$$

To calculate G (the shunt resistance between the signal conductor and the ground plane), the self-capacitance of the line is found using the equations from Section 2.3.

$$\text{TD} = \frac{\sqrt{\varepsilon_r}}{c} = \frac{\sqrt{4}}{3 \times 10^8 \text{ m/s}} = 6.67 \text{ ns/m}$$

$$Z_o = \frac{60}{\sqrt{4.0}} \ln \frac{4(10 + 10 + 0.63)}{0.67\pi[0.63 + 0.8(5)]} = 64 \ \Omega$$

$$C = \frac{\text{TD}}{Z_o} = \frac{\sqrt{LC}}{\sqrt{L/C}} = \frac{6.67 \times 10^{-9}}{64} 104 \text{ pF/m.}$$

The equivalent resistance due to the finite conductivity of the dielectric is calculated using equation (4.17). The loss tangent at 400 MHz is found from Figure 4.11.

$$G = \frac{\varepsilon''}{\varepsilon'}(2\pi F C_{11}) = 0.013(2\pi)(400 \text{ MHz})(104 \text{ pF})$$

$$= 0.0034 \text{ m}/\Omega \Rightarrow \frac{1}{G} = 294 \ \Omega/\text{m}$$

Note that the total shunt resistance due to the dielectric losses is almost 15 times larger than the series resistance of the conductor.

4.2. VARIATIONS IN THE DIELECTRIC CONSTANT

The dielectric constant of the PCB substrate, ε_r, directly affects the signal transmission characteristics of high-speed interconnects. Some of the characteristics that are dependent on ε_r include propagation velocity, characteristic impedance, and crosstalk. The value of ε_r is not always constant for a given material but varies as a function of frequency, temperature, and moisture absorption. Additionally, for a composite material, the material dielectric properties will change as a function of the relative proportions of its components [Mumby, 1988].

The substrate of choice for most PCBs in commercial applications is a composite called FR4, which consists of an epoxy matrix reinforced by a woven glass cloth. This composite features a wide range of thickness and relative compositions of glass and resin. Consequently, the dielectric properties observed for FR4 laminates can differ substantially from sample to sample. Most vendors will provide a value of ε_r for only one frequency. To ensure a robust digital design that will yield adequate performance over all manufacturing and environmental tolerances, it is important to consider the variations in the dielectric constant.

A first-order approximation of the dielectric constant of FR4 composite may be calculated as

$$\varepsilon_r = \varepsilon_{\mathrm{rsn}} V_{\mathrm{rsn}} + \varepsilon_{\mathrm{gls}} V_{\mathrm{gls}} \qquad (4.18)$$

where $\varepsilon_{\mathrm{rsn}}$ and $\varepsilon_{\mathrm{gls}}$ is the dielectric constants of the epoxy resin and the woven glass cloth and V_{rsn} and V_{gls} are the volume fractions of the resin and glass cloth. The relative volume fractions of glass to resin can change from sample to sample. Subsequently, it is possible that different layers of a PCB board can be manufactured with samples from different lots of FR4. Subsequently, relatively large variations in the dielectric constant are possible between layers on the same PCB. Measurements have also shown that the dielectric constant of FR4 can vary with frequency and resin content. A typical plot of dielectric constant variations versus frequency for a volume resin fraction of 0.724 is shown in Figure 4.12.

An approximate equation for the prediction of the relative dielectric constant of FR4 as a function of frequency and resin content is

$$\varepsilon_r(V_{\mathrm{rsn}}, F) = 6.32 - [2.17 + 0.168 \log_{10} F(\mathrm{kHz})] V_{\mathrm{rsn}} \qquad (4.19)$$

This relationship is supported by extensive data acquired for FR4 laminates and has been shown to be accurate within a few percent of experimental measurement over the range 1 kHz to 1 GHz. Since this equation was derived empirically, caution should be taken when extrapolating to frequencies above 1 GHz [Mumby, 1988]. It has been determined through experimental measurement that the glass cloth reinforcement experiences no dielectric constant

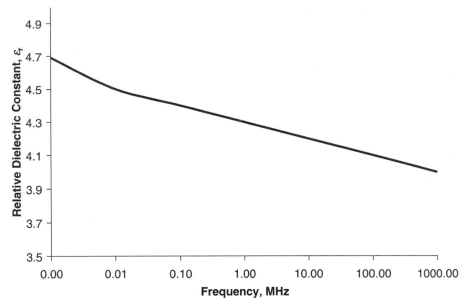

FIGURE 4.12   Dielectric variation with frequency for a typical sample of FR4. (Adapted from Mumby [1988].)

variation in this frequency range; subsequently, the equation is only a function of the frequency and the resin content.

The PCB manufacturer should be able to provide the volume content of the resin. However, if this information is not attainable, it can be estimated as

$$V_{rsn} \approx 1 - \frac{H_{gls}}{H} \tag{4.20}$$

where H_{gls} is the total thickness of the glass cloth and H is the total thickness of the dielectric layer. When PCBs are manufactured, the dielectric layers are built up to the desired thickness by stacking several layers of glass cloth and gluing everything together with the epoxy resin. The manufacturer can provide the type and thickness of glass cloth used and the number of layers used, which will yield H_{gls}. The total layer thickness can either be measured via cross-sectioning techniques or obtained from the manufacturer.

4.3. SERPENTINE TRACES

When a layout engineer is routing the board, it is usually impossible to route each net in a perfectly straight line. Board aspect ratios, timing requirements, and real estate limitations inevitably require that the traces be routed in serpentine patterns such as depicted in Figure 4.13. Extensive serpentine patterns

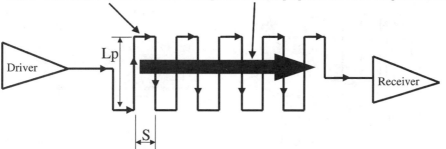

FIGURE 4.13 Example of a serpentine trace.

will probably be encountered when the design specification of a digital system requires all the traces on the PCB to be length equalized and there is limited real estate on the board to do so. Serpentine traces are also often used to fix timing "hold" problems by delaying the data with respect to the clock.

The effect of a serpentine trace is seen in both the effective propagation delay and the signal integrity. These effects are caused primarily by self-coupling between the parallel sections of transmission line (L_p in Figure 4.13). To understand this, imagine a signal propagating down a transmission line. If the trace is serpentined with enough space between parallel sections to eliminate crosstalk effects, the waveform seen at the receiver will behave as if it were propagating down a straight line. However, if significant crosstalk exists between the parallel sections, a portion of the signal will propagate in a path that is perpendicular to the serpentine via the mutual inductance and capacitance as depicted in Figure 4.13. Subsequently, a component of the signal will arrive early, which will affect the signal integrity and the delay. Figure 4.14 shows the difference between a 5-in. straight line and a 5-in. serpentine line with 5- and 15-mil spacing. Notice that as the space between the parallel sections is increased (S in Figures 4.13 and 4.14), the waveform approaches the ideal straight-line case. These simulations were performed on buried microstrip lines with a relative dielectric constant of 4.2, which produced a propagation delay of approximately 174 ps/in. (see Chapter 2). Therefore, if no coupling were present between the parallel sections, the signal would arrive at the receiver in approximately 870 ps for a 5-in. trace. Figure 4.14, however, shows components of the signal arriving much earlier than the ideal 870 ps for the serpentined traces. These early components are the portion of the signal that travels perpendicular to the coupled parallel section via the mutual parasitics. Notice that the magnitude of the ledges change significantly for differing degrees of coupling, but the time duration of the ledges remain constant. The duration of the ledges are proportional to the physical length of the coupled sections (L_p) and the voltage magnitude of the ledges depend on the space between parallel sections.

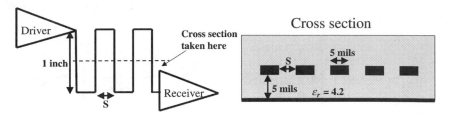

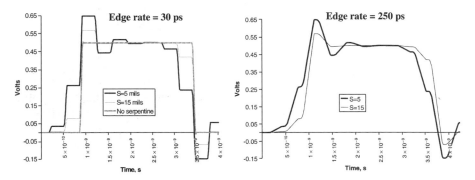

FIGURE 4.14   Effect of a serpentine trace on signal integrity and timing.

It should be noted that even if the signal integrity impact does not cause a timing problem directly (i.e., the ledges occur outside the threshold region), it can contribute to other problems, such as ISI (which is discussed in Section 4.4).

RULE OF THUMB: Serpentine Traces

The following guidelines will help minimize the effect of serpentine traces on signal integrity and timing:

- Make the minimum spacing between parallel sections (S) at least $3H$ to $4H$, where H is the height of the signal conductor above the reference ground plane. This will minimize coupling between parallel sections.
- Minimize the length of the serpentined sections (L_p) as much as possible. This will reduce the total magnitude of the coupling.
- Embedded microstrips and striplines exhibit fewer serpentine effects than do microstrip lines.
- Do not serpentine clock traces.

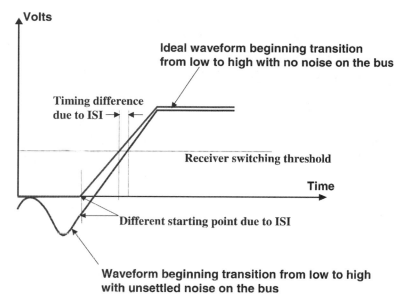

FIGURE 4.15   Effect of ISI on timings.

4.4. INTERSYMBOL INTERFERENCE

When a signal is transmitted down a transmission line and the noise on the bus due to reflections, crosstalk, or any other source has not settled completely, the signal launched onto the line will be affected, degrading both the timing and the signal integrity margins. This is referred to as *intersymbol interference* (ISI) *noise*. ISI is a major concern in any high-speed design, but especially so when the period is smaller than two times the delay of the transmission line. ISI must be analyzed rigorously in system design because it is often a dominant effect on performance. Figure 4.15 shows a graphical example of how ISI can affect timings. Note the timing difference between the ideal waveform and the noisy waveform that begins the transition with unsettled noise on the bus. Left unchecked, this timing difference can be several hundred picoseconds and can consume all available timing margins in a high-speed design.

To capture the full effects of ISI, it is important to perform many simulations with long pseudorandom bit patterns, and the timings should be taken at each transition. The bit patterns should be chosen so that all system resonances are sufficiently excited and the noise is allowed to settle partially prior to the next transition. To capture most of the timing impacts, however, simulations can be performed with a single periodic bit pattern at the fastest bus period and then at $2\times$ and $3\times$ multiples of the fastest bus period. For example, if the fastest frequency the bus will operate at is 400 MHz, the pulse duration of a single bit will be 1.25 ns. The data pattern should be repeated with pulse durations of 2.5 and 3.75 ns. This will represent the following data patterns transitioning at the highest bus rate:

010101010101010
001100110011001
000111000111000

The maximum difference in flight time, or flight-time skew between these patterns, produces a first-order approximation of the ISI impact (Chapter 9 defines flight time as used in digital system design). This analysis can be completed in a fraction of the time that it takes to perform a similar analysis using long pseudorandom patterns. It must be stressed, however, that the most accurate impact of ISI must be evaluated with long pseudorandom bit patterns.

ISI will also dramatically affect the signal integrity. Figure 4.16 shows the effect that different bit patterns have on the wave shape in a system with significant reflections. It is important to investigate different bit patterns to ensure

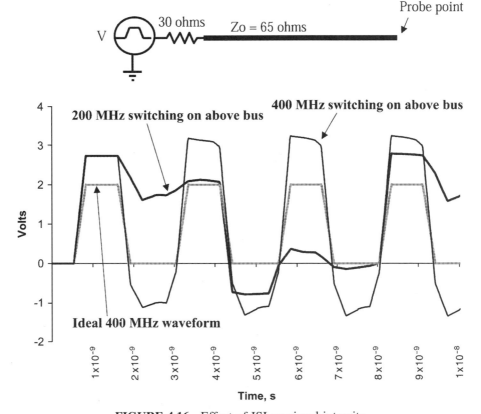

FIGURE 4.16   Effect of ISI on signal integrity.

a robust design. If a signal is switching at the same time that ringback is occurring, for example, the ringback will be masked at one switching rate but will not be masked for other switching rates, as shown in Figure 4.16. Figure 4.16 is a very simple example that demonstrates how dramatically the bit pattern can affect the signal integrity. Realistic bus topologies, especially those with multiple loads, will have very complicated ISI dependencies that can devastate a design if not accounted for. In Chapter 9 we describe methodologies that can account properly for ISI in the design process.

RULE OF THUMB: ISI

The following guidelines will help to minimize the effect of ISI.

- Minimize reflections on the bus by avoiding impedance discontinuities and minimizing stub lengths and large parasitics (i.e., from packages, sockets, or connectors).
- Keep interconnects as short as possible.
- Avoid tightly coupled serpentine traces.
- Avoid line lengths that cause signal integrity problems (i.e., ringback, ledges, overshoot) to occur at the same time that the bus can transition.
- Minimize crosstalk effects.

4.5. EFFECTS OF 90° BENDS

Virtually every PCB design will exhibit bends in some or all the traces. Thus, how to account for a bend in a transmission line may be important to a simulation. When considering how to model a bend, several things may be considered that may add unnecessary complexity to a model. It has been demonstrated through correlation to empirical measurements that a simple lumped-capacitance model is adequate for most systems. However, the reader should be aware of the limitations of this model to realize when and if the model needs to be revised.

The empirically inspired model for a 90° bend is simply one square of excess capacitance in the transmission line model shown in Figure 4.17. This means a capacitance with a value equal to a segment of transmission line equal to its width. This capacitance should be added to a simulated transmission line at the point where the bend occurs. The excess capacitance for a 90° bend for a typical 50- to 65-Ω line widths is approximated as

$$C_{90° \text{ bend}} \approx C_{11}w \tag{4.21}$$

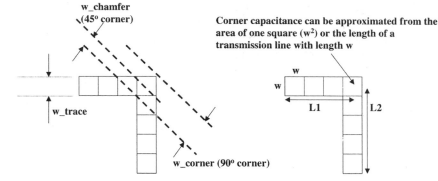

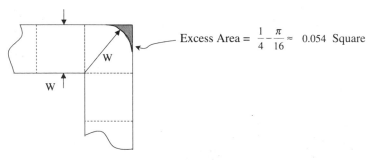

FIGURE 4.17   Extra capacitance from a 90° bend.

where C_{11} is the self-capacitance of the line and w is the line width. Although this excess capacitance is typically very small, it can cause problems with wide traces and with large numbers of bends. If this small amount of excess capacitance is a concern, simply rounding the corners to produce a constant width around the bend will virtually eliminate the effect. Round corners, however, cause problems with many layout tools. Another approach is to chamfer the edge by 45°. The simplest method, however, is to completely avoid the use of 90° bends by using 45° bends instead. The excess capacitance of a 45° bend is significantly smaller than a 90° bend and can be ignored for most applications.

The reader may have noted in equation (4.21) that the extra capacitance of one full square is more than one might expect if one were to consider the extra capacitance from the extra area as shown in Figure 4.18. The reason that empirical measurements may have favored one full square of capacitance rather than the smaller area capacitance shown in Figure 4.18 is not clear.

FIGURE 4.18   Excess area of bend. The excess area is less than the 1 square of empirically inspired excess capacitance for a 90° bend.

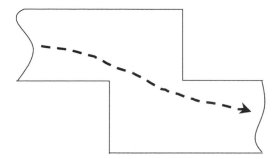

FIGURE 4.19 Some component of the current may hug corners leading to signals arriving early at destination.

There is one more empirically conjectured effect that is worth noting in this section. Some components of the current flow in a transmission line will flow in such a manner that they will deviate from the expected delay based on the layout length. In Figure 4.19, consider the dashed line, which might be a component of the current. Since the current cut both corners, that component of the current will arrive at the destination slightly earlier than expected. Thus, in a system with many bends, the delay may be slightly different than expected. This effect has been seen in laboratory measurements and is mentioned only as a precaution to keep in mind, particularly when two or more separately routed lines with many bends must be of equal delay.

4.6. EFFECT OF TOPOLOGY

So far in this book we have covered many issues that deal with an interconnect connecting two components. However, this is not always the case. Often, it is required that a single driver be connected to two or more receivers. In these cases the topology of the interconnects can affect the system performance dramatically. For example, consider Figure 4.20, which is a case where one driver is connected to two receivers. In one case, the impedance of the base (Z_{o1}) is equal to the impedance of the two legs (Z_{o2}). When the signal propagates to the junction, it will see an effective impedance of $Z_{o2}/2$, resulting in the waveform in the figure that steps up toward the final value. These reflections can be calculated using a lattice diagram similar to that of Figure 2.17.

When the impedance of the legs are twice the impedance of the base, the effective impedance the signal will see at the junction will be equal to the base; subsequently, there will be no reflections generated. This case, when $Z_{o2} = 2Z_{o1}$, is shown in Figure 4.20. When the structure is unbalanced, as in the case where one leg is longer than the other, the signal integrity will deteriorate dramatically because the reflections will arrive at the junction at different times. This is shown in Figure 4.21. The signal integrity can be

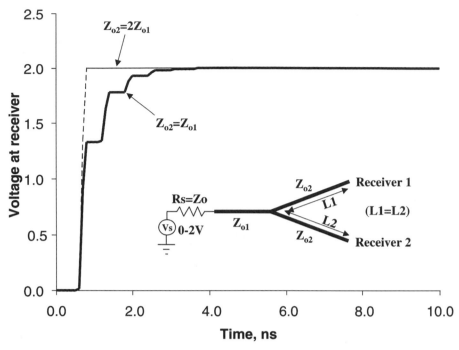

FIGURE 4.20 Signal integrity produced by a balanced T topology.

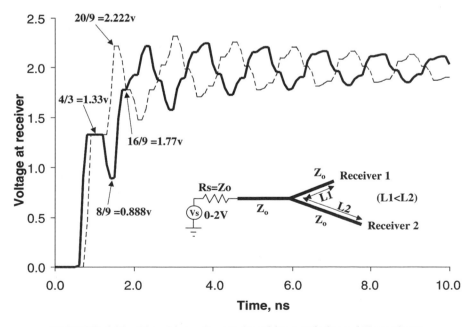

FIGURE 4.21 Signal integrity produced by a unbalanced T topology.

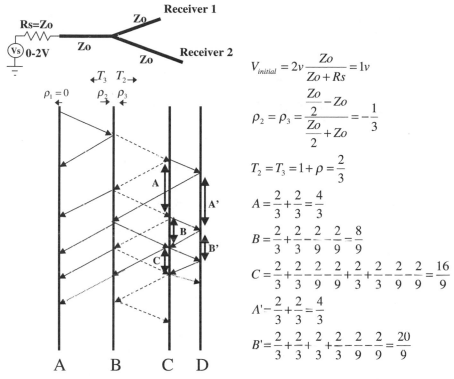

$$V_{initial} = 2v \frac{Zo}{Zo + Rs} = 1v$$

$$\rho_2 = \rho_3 = \frac{\frac{Zo}{2} - Zo}{\frac{Zo}{2} + Zo} = -\frac{1}{3}$$

$$T_2 = T_3 = 1 + \rho = \frac{2}{3}$$

$$A = \frac{2}{3} + \frac{2}{3} = \frac{4}{3}$$

$$B = \frac{2}{3} + \frac{2}{3} - \frac{2}{9} - \frac{2}{9} = \frac{8}{9}$$

$$C = \frac{2}{3} + \frac{2}{3} - \frac{2}{9} - \frac{2}{9} + \frac{2}{3} + \frac{2}{3} - \frac{2}{9} - \frac{2}{9} = \frac{16}{9}$$

$$A' = \frac{2}{3} + \frac{2}{3} = \frac{4}{3}$$

$$B' = \frac{2}{3} + \frac{2}{3} + \frac{2}{3} + \frac{2}{3} - \frac{2}{9} - \frac{2}{9} = \frac{20}{9}$$

FIGURE 4.22   Lattice diagram of an unbalanced T topology.

calculated with the use of a lattice diagram, although only a masochist would attempt to do so instead of simulating it with a computer. Figure 4.22 shows a lattice diagram that calculates the first few reflections in Figure 4.21 (the author must be a masochist). Referring to the figure, vertical lines *A* and *B* represent the electrical pathway between the driver and the junction, lines *B* and *C* represent the path between the junction and receiver 1, and lines *B* and *D* represent the pathway between the junction and receiver 2. As more legs are added to the topology, it becomes ever more sensitive to differences in the electrical length of the legs. Furthermore, differences in the loading at each leg will cause similar instabilities.

So what can we learn from this? The answer is symmetry. Whenever a topology is considered, the primary area of concern is symmetry. Make certain that the topology looks symmetric from the point of view of any driving agent. This is usually accomplished by ensuring that the length and the loading are identical for each leg of the topology. The secondary concern in to try and ensure that the impedance discontinuities at the topology junctions are minimized, although this may be impossible in some designs.

CHAPTER 5

Connectors, Packages, and Vias

So far in this book we have covered most of the issues associated with the PCB board. The basic fundamentals of a transmission line were explored in Chapter 2, Chapter 3 dealt with crosstalk, and in Chapter 4 we explained many of the nonideal transmission line issues with which modern designers must contend. Subsequently, we have explored in detail many of the issues associated with the electrical pathway from the pin of the driver to the pin of the receiver. The silicon, however, is located in a package, and the signal usually transverses through vias and connectors. In this chapter we explain principles that will allow the designer to extend analysis to account for the entire pathway, from the silicon pad at the driver to the silicon pad at the receiver, by exploring packages, vias, and connectors.

5.1. VIAS

A via is a small hole drilled through a PCB that is used to make connections between various layers of the PCB or to connect components to traces. It consists of the barrel, the pad, and the antipad. The *barrel* is a conductive material that fills the hole to allow an electrical connection between layers, the *pad* is used to connect the barrel to the component or trace, and the *antipad* is a clearance hole between the pad and the metal on a layer to which no connection is required. The most common type of via is called a *through-hole via* because it is made by drilling a hole through the board, filling it with solder, and making connections on appropriate layers via the pad. Other, less common types of vias, used primarily in multichip modules (MCMs) and advanced PCBs, are blind, buried, and micro-vias. Figure 5.1 depicts a typical through-hole via and its equivalent circuit. Notice that the pads used to connect the traces on layers 1 and 2 make contact with the barrel and that there is no connection on layer 3. Blind and buried vias have a slightly different construction. Since through-hole vias are by far the most common used in industry, they are the focus of this discussion.

Notice that the via model is simply a pi network. The capacitors represent the via pad capacitance on layers 1 and 2. The series inductance represents the barrel. Since the via structures are so small, they can be modeled as lumped

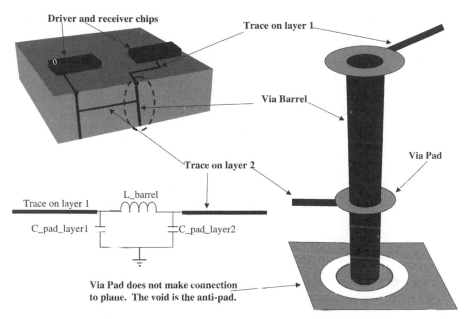

FIGURE 5.1 Equivalent circuit of a through-hold via.

elements. This assumption, of course, will break down when the delay of the via is larger than one-tenth of the edge rate. The main effect that via capacitance has on a signal is that it will slow down the signal edge rate, especially after several transitions. The amount that the signal edge rate will be slowed can be estimated by examining the degradation of a signal transmitted through a capacitive load, as shown later in this chapter in equation (5.21). Furthermore, if several consecutive vias are placed in close proximity to one another, it will lower the effective characteristic impedance, as explained in Section 5.3.3. The approximate value of the pad capacitance is [Johnson and Graham, 1993]

$$C_{\text{via}} \approx \frac{1.41 \varepsilon_r D_1 T}{D_2 - D_1} \qquad \text{picofarads} \qquad (5.1)$$

where D_2 is the diameter of the antipad, D_1 the diameter of the via pad, T the thickness of the PCB, and ϵ_r the relative dielectric constant. The typical total capacitance of a through-hole via is approximately 0.3 pF. It should be noted that the via model depicted in Figure 5.1 assumes that half of this total capacitance is attributed to each pad.

The inductance of the vias is usually more important than the capacitance to digital designers. The vias will add a small amount of series inductance to the system, which will degrade the signal integrity and decrease the effect of decoupling capacitors. The via inductance is modeled by a series inductor as shown in Figure 5.1 and can be approximated by [Johnson and Graham, 1993]

$$L \approx 5.08h \left[\ln \left(\frac{4h}{d} \right) + 1 \right] \qquad \text{nanohenries} \qquad (5.2)$$

where h is the via length and d is the barrel diameter.

5.2. CONNECTORS

Connectors are devices that are used to connect one PCB board to another. As speeds increase, connector design becomes increasingly difficult. A good example of a high-speed connector in a modern digital system is the slot 1 connector used to connect the Pentium III cartridge processor to the motherboard. Another, more advanced example is the RIMM (Rambus Inline Memory Module) connector, which is operating at speeds of 800 megatransfers per second.

Since the geometries of connectors are usually complex, it is impossible to accurately calculate the effective parasitics without the assistance of a two- or a three-dimensional field solver or measurements. However, the fundamental effects and a basic understanding of how a connector will affect system performance can be learned early in the design process by examining a first-order-level model.

In this section we examine the properties important to the design and modeling of high-speed connectors by concentrating on the fundamental issues. The main topics explored in this section are connector crosstalk, series parasitics, and return current path inductance. Figure 5.2 depicts a conceptual connector that will be used to demonstrate the detrimental effects on signal integrity.

5.2.1. Series Inductance

The most fundamental effect of a connector is the addition of a series inductance, which can be estimated to be a first-order level model with the use of simple straight-wire formulas. Estimates for the series inductance of round and rectangular wire, respectively, are [Poon, 1995]

$$L \approx \frac{\mu_0}{2\pi} l \left[\ln \left(\frac{2l}{r} \right) - \frac{3}{4} \right] \qquad \text{nanohenries} \qquad \text{for } r \ll 1 \qquad (5.3)$$

$$L \approx \frac{\mu_0}{2\pi} l \left[\ln \left(\frac{4l}{p} \right) + \frac{1}{2} \right] \qquad \text{nanohenries} \qquad (5.4)$$

where μ_0 is the permeability of free space, l the length, r the radius of the wire, and p the perimeter. It should be noted that the length is the primary

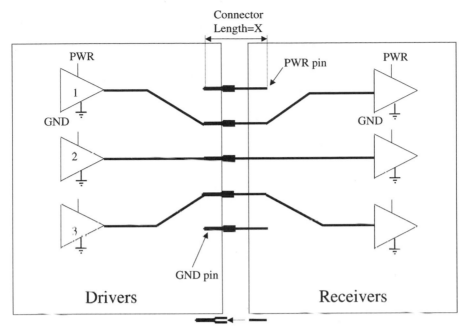

FIGURE 5.2   Example of a PCB connector.

contributor to inductance; the shape of the conductor is not terribly significant as long as the cross-sectional width is small compared to the length.

5.2.2. Shunt Capacitance

Although shunt and mutual capacitance is also an important effect to consider in connector design, it can usually be ignored in the initial "back of the envelope" estimations of connector performance. The main effect that capacitance will have on the system is that it will slow down the system edge rate. It should be noted that the addition of extra capacitance is sometimes used to mitigate the impedance discontinuity seen at the connector. The addition of capacitance will lower the effective impedance of the pin. This must be analyzed very carefully and rigorously to guarantee that the design is sound. Extra capacitance can be added by using a wider pad, adding a small tab or widening out the connector pins. It should be noted that it is difficult to estimate the effects of capacitance without the use of two- or three-dimensional tools or laboratory measurements.

5.2.3. Connector Crosstalk

Crosstalk also plays a significant role in connector performance. Typically, the mutual inductance plays a larger role than mutual capacitance. Subsequently, for first-order approximations, it is usually adequate to ignore the mutual capacitance. When more accurate modeling is necessary, both two- and three-dimensional simulators, or measurements, can be used to determine accurate mutual parasitics.

The mutual inductance between two connector pins can be can be approximated as [Poon, 1995]

$$L_M \approx \frac{\mu_0}{2\pi} l \left[\ln \left(\frac{l}{s} + \sqrt{1 + \left(\frac{l}{s}\right)^2} \right) - \sqrt{1 + \left(\frac{s}{l}\right)^2} + \frac{s}{l} \right] \qquad \text{nanohenries} \tag{5.5a}$$

$$L_M \approx \frac{\mu_0}{2\pi} l \left[\ln \left(\frac{2l}{s} \right) - 1 \right] \qquad \text{nanohenries} \qquad \text{for} \quad s \ll l \tag{5.5b}$$

where μ_0 is the permeability of free space, l the length, and s the center-to-center spacing between the wires. It should be noted that the mutual inductance is largely independent on the cross-sectional area and the shape of the conductor.

5.2.4. Effects of Inductively Coupled Connector Pin Fields

Consider the connector in Figure 5.2. The connector pins essentially form a network of coupled inductors. For the purposes of this exercise, we consider three drivers, one power pin, and one ground pin. The target will be net 2. The total induced voltage on pin 2 due to current changes in itself and pins 1 and 3 is

$$v_2 = L_{22}\dot{I}_2 + L_{21}\dot{I}_1 + L_{23}\dot{I}_3 \tag{5.6}$$

where $\dot{I}_n$ represents dI_n/dt. Assuming that the currents and the slew rates for each buffer are the same, a single-line equivalent model can be created to simplify the analysis (see Section 3.6.2):

$$v_2 = \begin{cases} (L_{22} + L_{21} + L_{23})\dot{I} & \text{(all bits switching in phase)} \tag{5.7} \\ (L_{22} - L_{21} - L_{23})\dot{I} & \text{(bit 2 switching out of phase} \\ & \text{with bits 1 and 3)} \tag{5.8} \end{cases}$$

Equations (5.6) through (5.8) demonstrate how multibit switching through the connector pins can cause pattern-dependent signal integrity problems by inducing inductive noise onto the nets. However, what is often neglected is the effect of the power and ground connection pins.

Consider the circuit shown in Figure 5.3. This is a typical configuration for a GTL + bus. When output switches low, the N device is turned on and the P device is turned off. A high transition occurs in the opposite manner. Let's

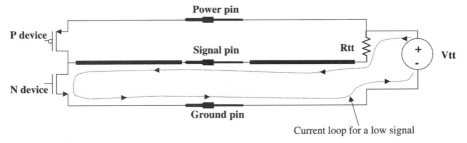

FIGURE 5.3 Current path in a connector when the driver switches low.

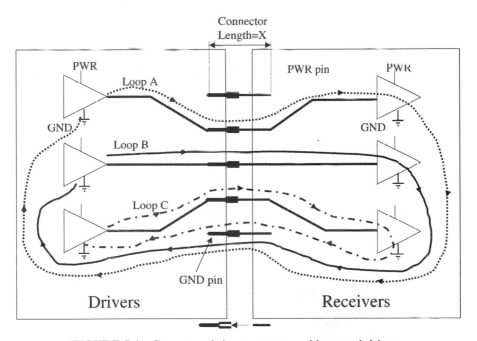

FIGURE 5.4 Current path in a connector with several drivers.

consider what the current does when the bus is pulled low by the N device. The current will be pulled out of V_{tt} and will travel through the transmission line, the signal pin, down through the N device, through the ground plane, through the ground pin, and back into the V_{tt} source. This is depicted by the current loop in Figure 5.3. Since a transient current will flow though the ground pin, it will induce inductive noise into the system of $L_{gnd}\dot{I}$. Subsequently, the inductance of the ground return path must be considered in the analysis. This effect is magnified significantly if the return current from several buffers shares the same ground pin, as depicted in Figure 5.4. In this particular case, the noise induced into the system would be $3L_{gnd}\dot{I}$, because three times as much current is flowing through the same ground pin. The same phenomenon holds true when transient current is flowing through the power pin. Care must be taken to understand how the return currents flow for the particular bus so that the effect of transient currents through the power and ground pins can be accounted for properly during the connector design. Various return current path scenarios for GTL and CMOS bus designs and their impact on the signal are explored in much greater detail in Chapter 6. It is imperative that the return current paths be understood so that the connector design can be optimized.

For the time being, let's assume that the return current will flow entirely back through the ground pin as depicted in Figures 5.3 and 5.4. To derive the effect of an inductance in the current return path, refer to Figure 5.5. Figure 5.5a represents a system of three inductive signal pins coupled to an inductive ground return pin. Note that all of the current flowing through the three signal inductors must return through the ground inductor. The effect of the ground return pin can be represented as shown in Figure 5.5b, assuming that the inductors are modified to include the inductive effects of the return path pin. This makes it easier to see the effect that an inductive pin in the ground return path has on the signal. The response of the system is shown in the following set of equations, which represent the response of Figure 5.5a:

$$V_2 = L_{22}\dot{I}_2 + L_{21}\dot{I}_1 + L_{23}\dot{I}_3 + L_{2g}\dot{I}_g$$

$$V_{gnd} = L_{gg}\dot{I}_g + L_{g1}\dot{I}_1 + L_{g2}\dot{I}_2 + L_{g3}\dot{I}_3$$

$$\dot{I}_g = -(\dot{I}_1 + \dot{I}_2 + \dot{I}_3)$$

$$V_2' = V_2 - V_{gnd}$$

$$V_2' = \dot{I}_1(L_{21} - L_{2g} - L_{g1} + L_{gg}) + \dot{I}_2(L_{22} - L_{2g} - L_{g2} + L_{gg})$$
$$+ \dot{I}_3(L_{23} - L_{2g} - L_{g3} + L_{gg})$$

(5.9)

V_2' is the voltage given by the simplified model. The result of (5.9) can easily be extended to a group of n conductors with a single current return path:

$$L_{ij}' = L_{ij} - L_{ig} - L_{gj} + L_{gg} \tag{5.10}$$

where the voltage V_2' in Figure 5.5b is given by

$$V_2' = L_{22}'\dot{I}_2 + L_{21}'\dot{I}_1 + L_{23}'\dot{I}_3 \tag{5.11}$$

Note that the effective inductance is simply the signal pin inductance, plus the ground return pin inductance, minus the mutual inductance.

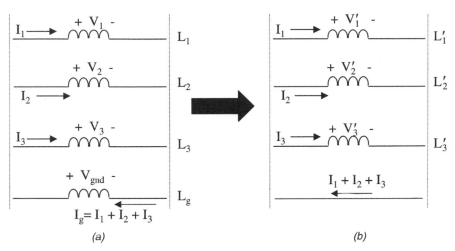

FIGURE 5.5 Incorporating return inductance into the signal conductor: (*a*) three inductive signal pins coupled to an inductive ground return pin; (*b*) effect of the ground return pin.

It should also be noted that the equations above are valid only for a coupled array of pins. The total return path inductance will increase with distance from the corresponding signal pin and should be modeled separately assuming that the path is significantly long or the total return path inductance is much greater than L_{gg}. The total current return inductance is the sum of the pin inductance and the inductance of the path to and from the return pin. The larger the total loop area in which the current flows, the larger the inductance. For example, the total loop inductance of loop *A* in Figure 5.4 has the largest total inductance, and loop *C* is the smallest.

Since the ground pins must return current to the power supply and the power pins must supply it to the drivers, low inductance is typically required for both power and ground paths to minimize inductive noise whether the return currents are flowing in the power or ground pins. Subsequently, it is generally optimal to maximize the total number of power and ground pins to decrease the total inductive path. In the equations above, it will effectively decrease L_{gg}. Furthermore, it is usually optimal to place power and ground pins adjacent to each other because the currents flows in opposite directions. Subsequently, the total inductance of the ground and power pins is reduced by the mutual inductance.

5.2.5. EMI

Another detrimental effect of a bad connector design is increased EMI radiation, covered in detail in Chapter 10. Notice the large current loops in Figure 5.4. As explained in detail in Chapter 10, the area of the loop is proportional to the emission radiated. Connectors will also exacerbate simultaneous switching noise by increasing the inductance in the ground return path and signal paths. Simultaneous switching noise and ground return path analysis are covered in detail in Chapter 6.

5.2.6. Connector Design Guidelines

Based on the analysis above, several general connector design guidelines can be made. The most obvious is to minimize the physical length of the connector pins to reduce the total series inductance. It is also desirable to maximize the power and ground to signal pin ratio. This will minimize the effect of the power and ground pin inductances. The placement of the power and ground pins should minimize the current loops and decrease connector-related radiated emissions. Each signal pin should be placed in close proximity to a power and a ground pin.

It should be noted that in cases where a large fraction of signals routed through a connector are differential, several of the design guidelines listed here will need modification. For instance, for connectors or packages that carry mostly differential signals, the number of power and ground pins required will probably be far less than those for a similar single-ended system. Furthermore, differential signals may sometimes be grouped together in large banks without power or ground pins separating them.

Several conclusions can be drawn from the discussion above.

1. Since coupling of signal pins to current return pins decreases the total inductance, it is optimal that each signal pin be tightly coupled to a current return pin (by placing them in close proximity to each other). The bus type and configuration will determine whether each signal should be coupled to both a power and a ground pin, just a ground pin, or just a power pin. In Chapter 6 we discuss how to determine where the return current is flowing. If the return current is flowing in the ground planes, the signal pins should be coupled to the ground pins. If the return current is flowing through the power planes, the signal pins should be coupled to the power pins. If the return current is flowing through both planes, each pin should, if possible, be coupled to both a power and a ground pin.

2. The power and ground pins should be placed adjacent to each other to minimize the inductance seen in the power and ground paths.

3. It is generally optimal to design a connector so that the ratio of power pins to signal pins and the ratio of ground pins to signal pins are greater or equal to 1. This will minimize the total return path inductance. This may not be as dominant a concern in differential systems.

4. It is generally optimal to use the shortest connector possible to minimize inductive effects and impedance discontinuities.

5. It is sometimes optimal to increase the connector pin capacitance to lower the impedance and minimize discontinuities. This can be achieved by widening the connector pins or by adding small tabs on the PCB.

6. Connector capacitance tends to slow the edge rates.

RULE OF THUMB: Connector Design

- Minimize the physical length of the connector pins.
- Maximize the ratio of power and ground pins to the signal pins. If possible, these ratios should be at least 1.
- Place each signal pin as close as possible to a current return pin.
- Place power pins adjacent to ground pins.

Example 5.1: Choosing a Connector Pin Pattern. Figure 5.6 depicts several pin-out options for an 8-bit connector. For the purpose of this exercise, assume that the return currents will be flowing equally through the power and ground pins. Option (*a*) will exhibit the worst performance because the power and ground pins are physically very far apart from the signal pins. This pin-out will maximize the inductive noise caused the power and ground return paths because the return currents flowing from each of the eight signals will

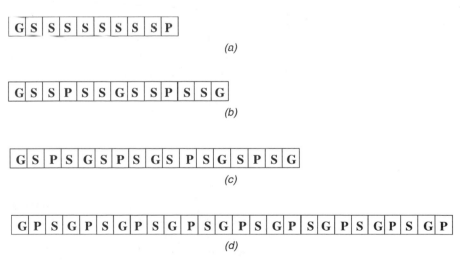

(a)

(b)

(c)

(d)

FIGURE 5.6 Eight-bit connector pin-out options assuming return current is flowing through both the power and ground pins: (*a*) inferior; (*b*) improved; (*c*) more improved; (*d*) optimal. G, ground pin; P, power pin; S, signal pin.

have to travel through a single power or ground pin, which is very far away. Furthermore, since the power and ground pins are physically very far away from the signals, large current loops can exacerbate connector-related EMI and increase series inductance. Finally, since all the signal pins are adjacent to each other, pin-to-pin crosstalk noise will be maximized, which will degrade signal integrity. The only benefit of this pin-out is that it will be physically as small as possible and very inexpensive to manufacture.

Option (*b*) is an improvement on option (*a*) because the power and ground pins are physically closer to the signal pins. Furthermore, there are an increased number of power and ground pins which provide for lower-inductance current return paths and provide better power delivery. Pin-to-pin crosstalk is also reduced because the power and ground pins will provide shielding between signal pairs. The crosstalk between pairs, however, can still be quite high. Subsequently, this connector is an ideal solution for a four-bit differential bus. Notice that the connector size has increased by 30% over option (*a*).

Option (*c*) is a further improvement. Notice that both a power and a ground pin surround each signal pin. This minimizes the return current path inductance through both the power and ground pins. Furthermore, the intermediate pins shield the signal pins from adjacent signal pins and dramatically reduce pin-to-pin crosstalk between signals. Notice, however, that the size of the connector has increased by 70% compared to option (*a*).

Option (*d*) is the optimal choice from a performance point of view (assuming that the return current is flowing in both the power and ground planes). Both a power and a ground pin surround each signal. Additionally, the power and ground pins are always adjacent to each other and the power/ground-to-signal ratio is 9 : 8. This minimizes the return current path inductance of both the power and ground pins, provides for significantly more shielding between the signal pins, and minimizes EMI effects. The obvious disadvantage of option (*d*) is that it is 2.6 times larger than option (*a*).

This example demonstrates that higher performance requires significantly larger connectors, which will affect the system in terms of cost and real estate. The designer must take great care in balancing performance with these constraints.

5.3. CHIP PACKAGES

A chip package is a holder for the integrated circuit. It provides the mechanical, thermal, and electrical connections necessary to properly interface the circuits to the system. There are many varieties of chip packages. The huge variety in packaging stems from the huge variety in system and circuit configurations. The number of different package types is growing constantly. Subsequently, it would be impossible to discuss each type of package that exists. In this section we concentrate on problems common to virtually all chip packaging schemes and discuss modeling options so that any type of package can be modeled. Package design must be optimized for the type of bus. We explore the basic attributes that govern the performance of a package for both a point-to-point bus such as the AGP (advanced graphics port on a PC) and a multidrop bus (such as a GTL front-side bus on a PC with more than one processor).

5.3.1. Common Types of Packages

The physical attributes of a package can typically be separated into three categories: the attachment of the die to the package, the on-package connections, and the attachment of the package to the PCB.

Attachment of the Die to the Package. Figure 5.7 depicts two of the most common methods of die attachment. Although many other techniques exist, these are by far the most common. First, let's explore the wire bond method of attaching the chip to the package. This is probably the most widely used method. A wire bond is simply a very small wire with a diameter on the order of 1 mil. Just for reference, the diameter of a human hair is approximately 3 mils. The primary impact of a bond wire is added series inductance. Bond wire lengths vary from approximately 50 to 500 mils. Since bond wires are so short, they are often modeled as a discrete inductor. There are several methods of obtaining the equivalent inductance of a bond wire. Equation (5.3) can be used as a quick approximation; however, the most accurate means

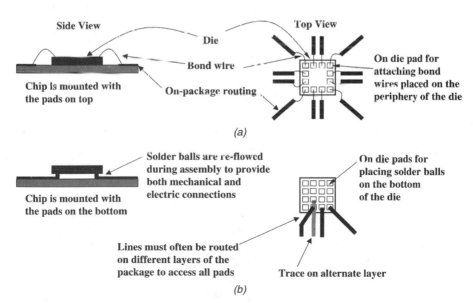

(a)

(b)

FIGURE 5.7 Common methods of die attachment: (*a*) wire bond attachment; (*b*) flip chip attachment.

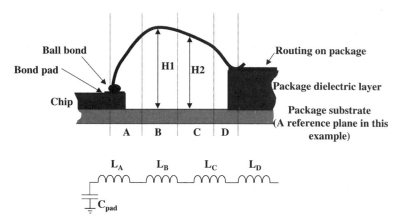

FIGURE 5.8   Modeling the arch of a bond wire.

of bond wire inductance calculation requires that the arch of the bond wire and the proximity of the ground plane be accounted for. If there is no local ground or power plane, equation (5.3) is an excellent approximation. However, if the wire passes over a plane, it is necessary to account for the arch and the height above the plane. Figure 5.8 depicts the best way to do this. This particular example divides the bond wire into four sections. The first section, A, is roughly perpendicular to the reference plane; subsequently, the reference plane will have a minimal effect and its contribution to the inductance can be calculated with equation (5.3). Section B is roughly parallel to the reference plane with an approximate height of H_1. Subsequently, the inductance L_B must be calculated using the inductance of a straight wire in the presence of a ground plane [Johnson and Graham, 1993]:

$$L = l(5.08 \times 10^{-9}) \ln \frac{4h}{d} \qquad \text{nanohenries} \qquad (5.12)$$

where l is the length in inches, h the height above the ground plane, and d the wire diameter. L_C must be calculated in the same manner as L_B using a height of H_2. L_D is calculated using equation (5.3) because it is roughly perpendicular to the reference plane. To gain better accuracy, a two-dimensional field simulator should be used to calculate the inductance of each section instead of the equation presented here. A three-dimensional simulator would allow a more exact inductance calculation; however, it is usually not worth the extra effort because the radius of the arch will change for each bond wire. Subsequently, any additional accuracy gained from the use of a three-dimensional simulator would be negated by the fact that it is impossible to quantify the exact physical characteristics of each bond wire in the system.

Wire bonds also exhibit a large amount of crosstalk, which will cause pattern-dependent inductance values and ground return path problems that will be governed by equations (5.6) through (5.11). If there is no local ground plane, equation (5.5) can be used to estimate the mutual inductance between two bond wires. Otherwise, the following equation should be used to approximate the effect of a local ground plane [Johnson and Graham, 1993]:

$$L_M = L \frac{1}{1 + (s/h)^2} \qquad \text{nanohenries} \qquad (5.13)$$

where L is the self-inductance of the two bond wires, s the center-to-center spacing, and h the height above the ground plane. A field solver should be used, however, to obtain the most accurate results. Furthermore, bond wires will tend to exacerbate rail collapse and simultaneous switching noise, which we explore in Chapter 6.

Although the use of bond wires leads to increased inductance and reduced signal integrity, the advantage is that they are inexpensive, mechanically simple, and allow for some changes in the bonding pad location and package routing. Furthermore, since the back of the chip is attached directly to the package substrate, it allows maximum surface area contact between the die and the package, which maximizes heat transfer out of the die. Additionally, when using bond wires, the I/O pads tend to be limited to the periphery of the die, which will inflate the die size for a large number of I/O.

Now let's consider flip-chip technology. Essentially, it is almost ideal from an electrical point of view. A flip-chip connection is obtained by placing small balls of solder on the pads of the die. The die is then placed upside down on the package substrate and the solder is re-flowed to make an electrical connection to the package bond pads. The pads are connected directly to the package interconnects, as shown in Figure 5.7. Flip-chip technology is also said to be *self-aligning* because when the solder is re-flowed, the surface tension of the solder balls will pull the die into alignment with the bond pads on the package.

The series inductance of a flip-chip connection is much lower than that of a wire bond. The typical inductance is on the order of 0.1 nH, which is an order of magnitude smaller than that of a typical bond wire. Furthermore, the effect of crosstalk in a flip-chip connection can be ignored. The bonding pads for a flip-chip connection can be placed over the entire die, not just on the periphery. This will help to minimize die size when a large number of I/O cells are required.

Mechanically and thermally, however, flip-chip technology is dismal. The thermal coefficient of expansion must be very close between the die and the package substrate. Otherwise, when the die heats up, it will expand at a different rate than the package, and the solder connections will be strained and can break. Furthermore, the physical tolerances must be very tight since the only degree of freedom when placing the chips is the small size of the pads. Cooling is also more difficult with flip-chip technology because the die is physically lifted off the package by the solder balls, which dramatically reduces the heat transfer and subsequently increases the cost of the thermal solution. Table 5.1 compares wire bonding and flip-chip technology.

TABLE 5.1. Comparison of Wire Bond and Flip-Chip Technology

	Series Inductance (nH)	Minimum Pitch (mils)	I/O Placement	Cooling
Wire bond	1–5	4–6	Periphery only	Easy
Flip chip	0.1	2–3	Entire surface	Difficult

Routing of the Signals on the Package. The routing of the signals inside the package typically fall into two categories; controlled impedance and non-controlled impedance. High-speed digital designs generally require the use of controlled-impedance packaging. A controlled-impedance package typically resembles a miniature PCB board with different layers and power and ground planes with a centered cavity used for die placement. The chip is usually attached with flip-chip bonding, although short wire bonds are often used. Small-dimensioned transmission lines are used to route the signals from the bond pad on the package to the board attachment. Controlled packages are significantly more costly than noncontrolled packages; however, they also provide for much higher system speeds. Noncontrolled-impedance packages usually wire bond the die directly to a lead frame, which is then soldered to the board. As one can imagine, this creates a series inductance that can wreak havoc on the signal integrity. Figure 5.9 depicts some generic examples of controlled- and noncontrolled-impedance packages.

Attachment of the Package to the Board. Attachment of the package to the PCB is achieved in a large variety of ways. Figure 5.9 depicts two popular methods, a lead frame (shown in the noncontrolled-impedance example) and a pin-grid array (shown in the controlled-impedance example).

A *lead frame* is simply a metal frame that is integral to the package, which provides an electrical connection between the wire bonds and the PCB. It is either attached with a through-hole mount or a surface mount. A through-hole mount is achieved by drilling a hole through the PCB, inserting the lead frame pins through the hole, and re-flowing solder. A surface-mount attachment is achieved by soldering the lead frame pins onto pads on the surface layer of the PCB. A *pin-grid array* (PGA) consists of an array of pins that stick out of the bottom part of the package. A good example of this is the familiar ceramic Pentium chip packages. Pin-grid arrays are often used with a socket because to allow the package to be removable.

There are other types of attachment techniques, such as a ball-grid array (BGA), which is attached to the PCB with an array of solder balls (a large version of flip-chip), and a land-grid array (LGA). Often there are two die mounted in a single package. The specific categories of packages are not covered in this book. A good overview of chip packages can be found in Section 2 of *Printed Circuits Handbook* [Coombs, 1996].

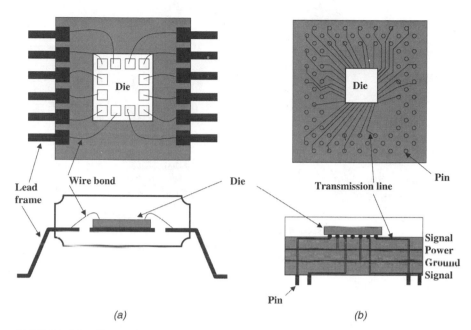

FIGURE 5.9 Comparison of (*a*) a noncontrolled and (*b*) a controlled impedance package.

5.3.2. Creating a Package Model

Now that each basic part of a package is understood, let's explore the process of creating a package model. For the time being, we focus on the signals. The issue of modeling the power supply path to the die is discussed in Chapter 6. As mentioned before, the most accurate way to model a package is to use field simulators. Depending on the geometry, it may or may not require full three-dimensional analysis. If the design is largely planar, a two-dimensional simulator will usually suffice. It is important for the reader to remember that all the modeling lessons learned in previous chapters apply.

1. Examine each stage of the package and determine the fundamental components of the model.

2. Perform simulations with field solvers to determine the parasitics of each section. Concentrate on groups of three to five lines so that the full effects of crosstalk can be accounted for.

3. Create distributed models of each section so that the model will behave correctly at the highest system frequency and at the highest system edge rate. If possible, use custom simulator elements such as HSPICE's W-Element to simplify the model and increase accuracy.

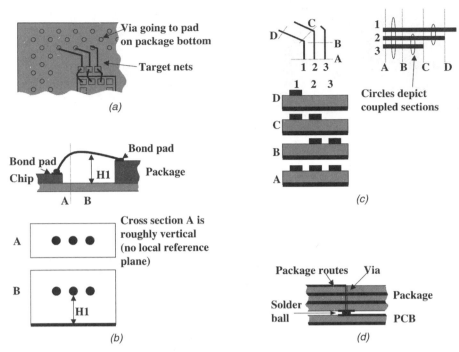

FIGURE 5.10 Components of a package: (*a*) modeling a BGA package; (*b*) attachment of die to package; (*c*) on-package routing; (*d*) connection to PCB.

Example 5.2: Creating an Equivalent Circuit Model of a Controlled-Impedance Package. Let's create an equivalent model of the controlled-impedance package in Figure 5.10. This particular package has short bond wires to connect the die to the package electrically, and the connection to the board is a ball-grid array. Tracing the signal path from the driver to the PCB allows us to determine the various parts of the package that need to be modeled. In this particular example, it is convenient to chop the problem into six sections: the silicon, the bond wires, the package routing, the via and ball pads, and the PCB traces. Usually, the capacitance of the on-die bond pads is provided by the I/O circuit designers. It is largely a function of the transistor gate sizes at the last stage of the output buffer and by the capacitance of the ESD (electrostatic discharge) protection diodes.

Attachment of the die to the package. Tracing the signal path from the silicon, the first section to model is the bond wire array. To model the bond wires correctly, they are split into two sections because the beginning portion is roughly vertical and the second portion is roughly parallel to the reference plane (see Figure 5.10*b*). The bond wire inductance can be estimated with the equations presented earlier, however, it is more accurate to use a two-dimensional simulator to calculate the parasitics of each section. The cross

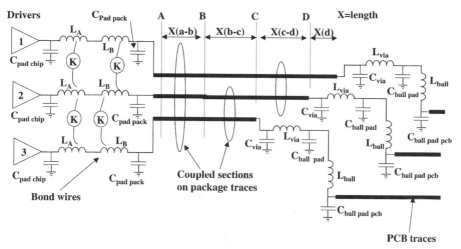

FIGURE 5.11 Equivalent circuit developed for Example 5.2

sections of each section are shown in Figure 5.10*b*. Once the mutual parasitics are known, the coupling factor, K, is calculated with equation (3.12). Coupling between bond wires should not be ignored because it can have significant impact. As a general rule of thumb, the inductance of a typical bond wire will be approximately 1 nH for each 50 mils of length. The capacitive parasitics of the bond wires have been ignored at this stage because they are extremely small and can usually be ignored. Although the power and ground bond wires are not included in this model, they can be dealt with in the same manner as power and ground pins in a connector array, as described earlier in this chapter. This is explored in more detail in Chapter 6. The equivalent circuit is depicted on the left hand side of Figure 5.11, where $C_{pad\ chip}$ is the I/O capacitance of the die, L_A and L_B are the bond wire inductance, and $C_{pad\ pack}$ represents the bonding pad on the package (where the bond wire attaches to the package).

Routing of the signals on the package. It is usually adequate to consider either three or five coupled signals on a package because coupling effects tend to reduce very quickly with distance. It should be noted, however, that densely routed traces tend to couple to multiple traces as they wind their way from driver to receiver. Subsequently, it is important to observe the package routing and determine which signals must be considered to achieve the nearest-neighbor coupling. In this particular example, trace 2 would be considered the target net, and lines 1 and 3 will be excited to account for crosstalk effects. The traces are separated into four sections represented by cross section *A* through *D* in Figure 5.10*c*. In section *A*, all three lines are coupled; in section *B*, only lines 2 and 3 are coupled; in section *C*, lines 1 and 2 are coupled; and finally, in section *D*, only line 1 continues because it is the longest. The package routing should entail all of the applicable effects outlined in Chapters 2 through 4, including losses. Losses are often ignored in package designs, which is a mistake because small dimensions tend to make them significantly more lossy than typical PCB traces. The models of the package routes are shown in the middle of Figure 5.11.

Attachment of the package to the PCB. This example is a BGA attachment. Subsequently, the major contributor to the performance will simply be the pad capacitance on the package and on the PCB. The via parasitics and the ball inductance will usually be small. The via parasitics can be estimated with equations (5.1) and (5.2). The ball inductance will usually be on the order of 0.5 nH. The only way to model the effect of the ball inductance rigorously is to use a three-dimensional simulator or to measure it. This section is shown on the right-hand side of Figure 5.11. C_{via} and L_{via} are the parasitics of the via that connects the package trace to the ball pad, $C_{\text{ball pad}}$ and L_{ball} are the parasitics of the solder ball and pad, and $C_{\text{ball pad PCB}}$ is the capacitance of the attachment pad on the PCB.

Example 5.3: Modeling a Pin-Grid-Array Attachment. A pin-grid array is a little easier to model but will still require the use of a full three-dimensional simulator to get the most accurate results. A good approximation of the pin inductance can be achieved with equations (5.3) and (5.5). To use these equations properly, the path of the current should be observed in detail. The total distance that the current travels in the pin should be used for the length. For an example, refer to Figure 5.12. The inductance $L_{\text{pin 1}}$ is approximated using length $X1$ and the radius of the pin. The inductance $L_{\text{pin 2}}$ is approximated using

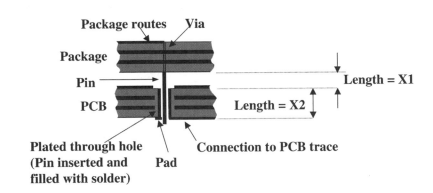

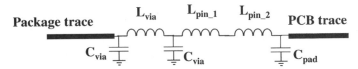

FIGURE 5.12 Modeling a pin grid array attachment.

length $X2$ and the radius of the plated through-hole. The entire radius of the plated hole is used because solder is re-flowed into the hole, making it a solid conductor.

5.3.3. Effects of a Package

The effect that a package has on system performance dependent on both the electrical attributes of the package and the type of bus in which it is used. In this section we explore the fundamental effects that a package will have on both a point-to-point and a multidrop bus. It should be noted that other effects, such as simultaneous switching noise, are covered in Chapter 6.

Point-to-Point Bus Topology. To demonstrate the effect of a package on a point-to-point topology, consider the following three cases depicted in the circuit in Figure 5.13.

- *Case 1*: noncontrolled-impedance package with a total bond length (bond wire plus lead frame) of 0.75 in.
- *Case 2*: controlled-impedance package with 0.25-in. bond wires and a 40-Ω transmission line on the package
- *Case 3*: controlled-impedance package with die flip-chip attached to the package and 40-Ω package routing

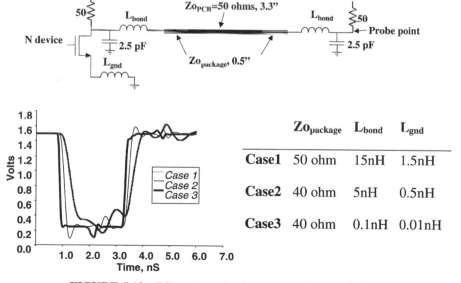

	ZO$_{package}$	L$_{bond}$	L$_{gnd}$
Case1	50 ohm	15nH	1.5nH
Case2	40 ohm	5nH	0.5nH
Case3	40 ohm	0.1nH	0.01nH

FIGURE 5.13 Effect of packaging on a point-to-point bus.

The test cases were chosen to be a simplified representation of a noncontrolled-impedance package, a controlled-impedance package with short bond wires, and a controlled-impedance package with flip-chip attachment on a basic GTL bus. Notice that the cleanest waveform is achieved with the flip-chip attachment (case 3). Since chip package routing is often lower in impedance than the motherboard (due to the manufacturable aspect ratios in many package designs), the package impedance was chosen to be 40 Ω. Case 2 is identical to case 3 except that the flip-chip attachment was replaced with 250-mil bond wires represented by the 5-nH inductors. Notice that the edge rate has been slowed significantly due to the extra inductance. This is because the inductance will act as a low-pass filter, which will attenuate the high-frequency components. Also note the decrease in the general signal quality due to the inductive noise. Finally, case 1 represents the noncontrolled impedance by eliminating the 40-Ω package section and replacing it with a 15-nH inductance. The 15 nH of inductance comprises the total inductance of a lead frame and a bond wire. Note that the edge rate has decayed even more significantly, and there is considerably more inductive noise to contend with. In a full system simulation, the package inductance can wreak havoc on signal integrity and timing budgets. Crosstalk between bond wires and leads cause significant pattern-dependent inductance values (due to the mutual inductance), which will cause edge-rate differences and degrade timing margins. Furthermore, crosstalk and inductive ground and power paths will cause simultaneous switching noise, as detailed in Chapter 6.

Multidrop or Daisy Chain Topologies. Multidrop bus topologies are very common. Two common examples are the front-side bus on a computer system connecting a chipset with multiple processors and RIMM memory modules. A block diagram of a multidrop bus is shown in Figure 5.14. The effect of packaging on such bus topologies is largely dependent on the package stub length. If the package stub is long, it will induce transmission line reflections that degrade the signal quality. If the package length is short, it will produce significant filtering effects that degrade the edge rate and change the effective characteristics of the transmission line.

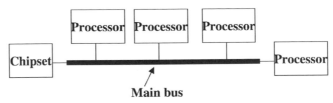

FIGURE 5.14 Example of a multidrop bus (front-side bus). Processors are tapped off the main bus.

Long Package Stub Effect. A package stub is considered to be long if it meets the following conditions:

$$\text{TD}_{\text{stub}} > \tfrac{1}{2}T_{10\text{–}90\%} \tag{5.14}$$

where TD_{stub} is the electrical delay of the unloaded package stub and $T_{10\text{–}90\%}$ is the rise or fall time of the signal edge.

Figure 5.15 shows a three-load topology driven from one end. Each capacitor represents the capacitance to the input stage of a receiver. The effect of package stub length and impedance is shown. Note that as the package stub length is increased, a ledge begins to form in the edge of the signal and increases the effective delay of the net (sometimes refered to as a

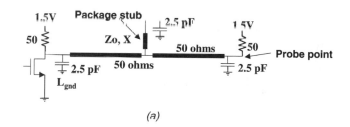

(a)

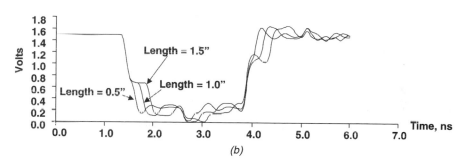

(b)

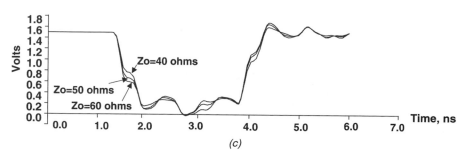

(c)

FIGURE 5.15 (*a*) Circuit; (*b*) effect of package stub length, *X*; (*c*) effect of package stub impedance.

timing push-out). The length of the ledge depends on the electrical delay of the stub. It should be noted that the capacitive loading at the end of the stub will increase the effective electrical delay of the stub by approximately 1 time constant (Z_oC). It should also be noted that the package stub does not induce a timing push-out in the form of a ledge if its delay is significantly smaller than the rise or fall times. This leads to a good rule of thumb which says that the electrical delay of the package stub, including the receiver capacitance, should be small compared to the rise or fall time.

Example 5.4: Calculating the Effect of a Long Package Stub.   The response of a system with a long stub is easy to calculate using simple techniques explored in Chapter 2. If the system has more than one stub, however, it is advisable to use a computer simulation. Figure 5.16 shows a simple system with one stub. The first reflection is calculated below. The entire response can be calculated using a lattice diagram. The reflection coefficient looking into the stub junction is calculated as

$$\rho_{\text{at stub}} = \frac{Z_s \| Z_{o_2} - Z_o}{Z_s \| Z_{o_2} + Z_o} = -\frac{1}{3} \tag{5.15}$$

The impedance looking into the junction will be a parallel combination of the stub impedance Z_s and the trunk impedance Z_{o_2}.

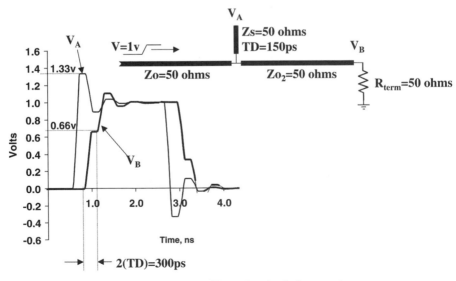

FIGURE 5.16   Effect of a single long stub.

The transmission coefficient down the stub and the second half of the trunk is shown as

$$T = 1 + \rho_{\text{at stub}} = 1 + \frac{Z_s \| Z_{o_2} - Z_o}{Z_s \| Z_{o_2} + Z_o} = \frac{2}{3} \tag{5.16}$$

Remember that current must be conserved; voltage does not. Subsequently, if the sum of the voltage transmitted down the stub and the second half of the trunk exceeds the input step voltage, don't be alarmed. Subsequently, the voltage from the first reflection at node B of Figure 5.16 is

$$V_B = VT = 1.0 \left(\tfrac{2}{3} \right) = 0.666 \text{ V}$$

The voltage from the first reflection at node A is

$$V_A = 2VT = 1.33 \text{ V}$$

which is doubled because there is no termination on the stub.

As shown in Figure 5.16, the timing push-out caused by the stub at node B depends on twice the total delay of the stub (the time for the signal to travel down the stub and for the reflection to come back).

Widely Spaced Short Package Stub Effect. A package stub is considered to be short if it meets the following conditions:

$$\text{TD}_{\text{stub}} < \tfrac{1}{2} T_{10\text{--}90\%} \tag{5.17}$$

where TD_{stub} is the electrical delay of the unloaded package stub and $T_{10\text{--}90\%}$ is the rise or fall time of the signal edge. When a package stub is sufficiently short, the load will look like a capacitor. The total capacitance will be the sum total of the I/O capacitance of the silicon and the capacitance of the stub. For a single capacitive load in the middle of a transmission line, such as that shown in Figure 5.17, the response can be calculated by plugging the parallel equivalent impedance of a capacitor, as calculated in equation (5.18),

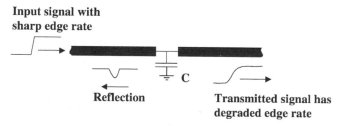

FIGURE 5.17   Effect of a short capacitive stub.

and inserted into equation (2.9).

$$Z_o \left\| \frac{1}{j\omega C} \right. = \frac{Z_o}{1 + Z_o(j\omega C)} \tag{5.18}$$

The reflection and transmission coefficients are

$$\rho_{cap} = \frac{Z_o \| (1/j\omega C) - Z_o}{Z_o \| (1/j\omega C) + Z_o} = \frac{-Z_o(j\omega C)}{2 + Z_o(j\omega C)} \approx \frac{-Z_o(j\omega C)}{2} \tag{5.19}$$

$$T_{cap} = 1 + \rho_{cap} \approx 1 - \frac{Z_o(j\omega C)}{2} \tag{5.20}$$

Note the form of equation (5.20). As the capacitance increases, the transmission coefficient decreases for a given frequency. This causes a low-pass filtering effect that will remove high-frequency components from the input signal. Subsequently, the signal transmitted through the shunt capacitor will have a degraded edge rate. Assuming a very fast input edge rate, the transmitted edge will have an edge rate approximately equal to 2.2 time constants (assuming 10 to 90% rise or fall times) between half the transmission line impedance and the value of the capacitance:

$$T_{edge} \approx \frac{2.2(Z_o C)}{2} \tag{5.21a}$$

As described in Appendix C, equation (5.21a) assumes that the input edge is a step function. To estimate the rise or fall time of the signal after it passes though the capacitive load, the equation

$$T_{system} \approx \sqrt{T_{edge}^2 + T_{input}^2} \tag{5.21b}$$

should be used, where T_{input} is the edge rate transmitted into the capacitive load.

Distributed Capacitive Loading on a Bus (Narrow Spacing). When the package stub length is short compared to the edge rates [so they look capacitive and conform to equation (5.17)] and loads are uniformly distributed so that the electrical delay between the loads is smaller that the signals rising or falling edges, the effect becomes distributed instead of lumped. A good example of this is a RIMM memory module. The loading essentially changes the characteristic impedance and the propagation velocity of the transmission line. Figure 5.18 depicts a distributed capacitive load due to several taps on a transmission line. The following equations are the equivalent impedance and delay parameters for a uniformly capacitive loaded

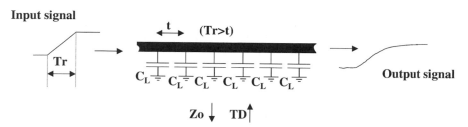

FIGURE 5.18 Effect of distributed short capacitive stubs.

bus:

$$Z'_o = \sqrt{\frac{L}{C + NC_L/X}} \qquad (5.22)$$

$$TD' = \sqrt{L\left(C + \frac{NC_L}{X}\right)} \qquad (5.23)$$

where L and C are the distributed transmission line parasitics as described in Chapter 2, C_L the total load capacitance (package stub + I/O), N the number of loads, and X the line length over which the loads are distributed. It should be noted that the capacitive loading will also have a significant low-pass filtering effect on the edge, as depicted in Figure 5.18. Several conditions other than packaging will cause this effect, such as narrowly spaced vias and 90° bends.

5.3.4. Optimal Pin-Outs

Optimal package pin-outs are chosen in largely the same manner as connector pin-outs, as described in Section 5.2. However, before a package pin-out is finalized, it is extremely important to perform a layout study to ensure that the pin-out chosen is routable. In some applications, pin-to-pin parasitic matching is more important than minimizing total pin parasitics. For instance, in applications in which pin-to-pin skew is a critical concern, a square package with pins on all four edges might be a better choice then a rectangular package with pins on only the two longest edges. This may be true even if some or all of the pins in the rectangular package may have smaller parasitic values. This is because the square package can be designed to take advantage of fact that its shape will allow less difference in lead frame length to the pins. This might lead to better performance even if the total parasitic per pin is greater for the square package then for several of the pins on the rectangular package, which is likely to have a larger relative difference in parasitics between the corner pin and a pin in the middle. This is illustrated in Figure 5.19. Similarly, when choosing pin-outs in a rectangular package, the pins closer to the middle of

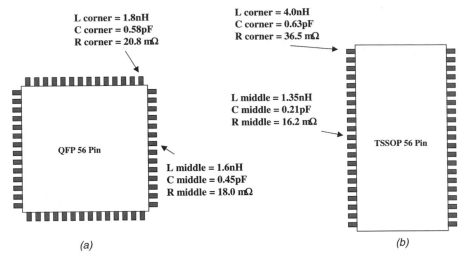

FIGURE 5.19 Package examples: (*a*) good pin-to-pin match; (*b*) low parasitics on best pins.

the package should be used for the highest-frequency or most critical signals, as these pins will have the least parasitic values.

RULES OF THUMB: Package Design

- Avoid the use of noncontrolled packaging for a high-speed design.
- Minimize inductance by using flip-chip or the shortest possible bond wire lengths for die attachment.
- Follow all the same pin-out rules of thumb as described for connector design when choosing power and ground pin ratios.
- Use the shortest possible package interconnect lengths to minimize impedance discontinuities and stub effects.
- On a multidrop bus, stub length is often the primary system performance inhibitor.
- In a multidrop bus design, never use long stubs that are narrowly spaced.
- Pin-to-pin package parasitic mismatch, which is strongly affected by the physical aspect ratio of the package, should be considered.

CHAPTER 6

Nonideal Return Paths, Simultaneous Switching Noise, and Power Delivery

The three subjects of this chapter—nonideal return paths, simultaneous switching noise, and local power delivery—are very closely related. In fact, sometimes the effects are indistinguishable in a laboratory setting. In the past, these effects were often ignored during the design process, only to be discovered and patched after the initial systems were built. As speeds increase, however, the patches become ineffective, and these effects can dominate the performance. The reader should pay close attention to this chapter because it is probably one of the most difficult subjects to understand. Furthermore, careful attention to the issues presented in this chapter is essential for any high-speed design. Often, designers hope that these issues will not be of consequence and subsequently ignore them. However, hope is a precious commodity. It is better to be prepared.

6.1. NONIDEAL CURRENT RETURN PATHS

So far, this book has covered virtually everything necessary to model the signal path as it propagates from the driving circuitry to the receiver through packages, sockets, connectors, motherboard traces, vias, and bends. Now it is time to focus on what is often considered the most difficult concept of high-speed design, nonideal return paths. Many of the effects detailed in this section are very difficult or impossible to model using conventional circuit simulators. Often, full-wave simulators are required to capture the full effects. Subsequently, in this section we focus less on specific modeling techniques and more on the general impact and physical mechanisms of the return current path that affect the system. As a general rule, great care should be taken to ensure that nonideal current return paths are minimized.

6.1.1. Path of Least Inductance

As was discussed in Chapter 2, the signal propagates between the signal trace and the reference plane. It does not merely propagate on the signal line. Subsequently, the physical characteristics of the reference plane are just as important

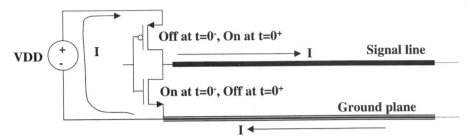

FIGURE 6.1   Return current path of a ground-referenced transmission line driven by a CMOS buffer.

as that of the signal trace. A very common mistake, even for experienced designers, is to focus on providing a very clean and controlled signal trace with no thought whatsoever of how the current is going to return. Remember that any current injected into a system must return to the source. It will do so through the path of least impedance, which in most cases means the path of least inductance. Figure 6.1 depicts a CMOS output buffer driving a microstrip line. The currents shown represent the instantaneous currents that occur when the driver switches from a low state to a high state. Just prior to the transaction (time = 0^-), the line is grounded via the NMOS. Immediately after the transaction (time = 0^+), the buffer switches to a high state and current flows into the line until it is charged up to V_{dd}. As the current propagates down the line, a mirror current is induced on the reference plane, which flows in the opposite direction. To complete the loop, the current must find the path of least inductance, which in this case is the voltage supply V_{dd}.

When a discontinuity exists in the return path, the area of the loop increases because the current must flow around the discontinuity. An increase in the current loop area leads to an increase in inductance, which degrades the signal integrity. Subsequently, the most fundamental effect of a discontinuity in the return path is an effective increase in series inductance. The magnitude of the inductance depends on the distance the current must diverge.

6.1.2. Signals Traversing a Ground Gap

To demonstrate the effect of a long-return current path, refer to Figure 6.2, which depicts a microstrip line traversing a gap in its ground reference plane. This is a convenient ground return path discontinuity to begin with because it is a simple structure, the return current path is well understood, and its trends can be applied to more general structures. As the signal current travels down the line, the return current is induced onto the ground plane. However, when the signal reaches the gap, a small portion of the ground current propagates across the gap via the gap capacitance, and the other portion is forced to travel around the gap. If there were no gap capacitance and the gap was infinitely long, the line crossing the gap would appear to behave as an open circuit, due

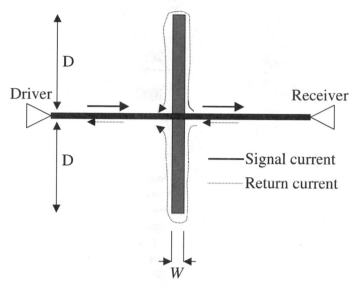

FIGURE 6.2 Driving and return currents when a signal passes over a gap in the ground plane.

to the impedance increase from the simultaneous increase in series inductance and decrease in capacitance to ground.

Figure 6.3 shows some general receiver wave-shape characteristics when a signal line crosses a gap, as in Figure 6.2. If the distance of the current return path (2D in Figure 6.2) is small compared to the edge rate, the gap will simply look like a series inductance in the middle of the transmission line. The extra inductance will filter out some of the high-frequency components of the signal and degrade the edge rate and round the corners. This is shown in the first graph of Figure 6.3a. When the electrical length of the return path, however, becomes longer than the rise or fall times, a ledge will appear in the waveform. The length of the ledge (in time) will depend on the distance the return current must travel around the discontinuity, as shown in Figure 6.3a. Since the width of the gap will govern the bridging capacitance, and thus the portion of current shunted across the gap $[I = C_{gap}(dV/dt)]$, the height of the ledge will depend on the gap width, as shown in Figure 6.3b. The larger the gap width, the less the capacitive coupling and the lower the height of the ledge.

Another effect of this specific nonideal return path is a very high coupling coefficient between traces traversing the same gap. The coupling mechanism is the gap itself. Energy is coupled into the slot and it propagates down to the other line via the slotline mode. A *slotline* is a transmission line in which the fields establish themselves between the conductors on both sides of the slot [Liaw and Merkelow, 1996]. The top portion of Figure 6.4 is an example

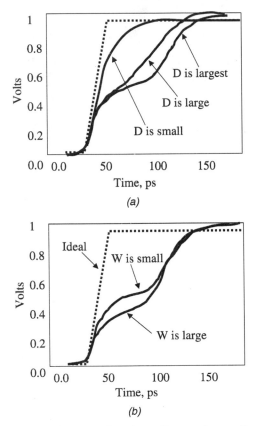

FIGURE 6.3 Signal integrity as a function of ground gap dimensions: (*a*) signal integrity as a function of return path divergence length; (*b*) signal integrity as a function of gap width. (Adapted from Byers et al. [1999].)

of coupling through a gap in the reference plane. Measurements will show significant coupling to other lines traversing the same gap even when the line is well over an inch away! In this case, nearly 15% of the total energy driven into the driver line is coupled to the adjacent line, 1.4 in. away [Byers et al., 1999].

Figure 6.4 also shows the TDR response of a line traversing a gap. Notice that from the driver's point of view, the return path discontinuity looks like a series inductance. Finally, the bottom portion of Figure 6.4 shows the response at the end of the line. Notice that when the return path distance (*D*) is small, the edge rate is decayed due to the inductive filtering effect. When *D* become larger, a ledge begins to form.

Modeling a Signal Traversing a Gap. For small current diversion paths (i.e., small gap lengths), the discontinuity can be modeled as a series inductor.

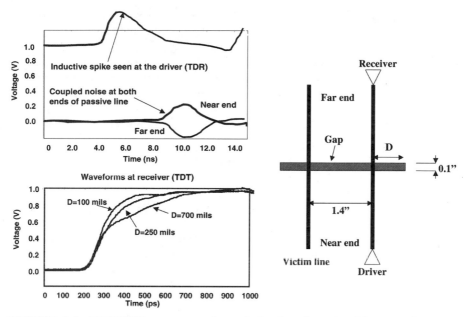

FIGURE 6.4 TDR/TDT response and coupled noise of a pair of lines passing over a slot in the reference plane.

Equation (6.1) and Figure 6.2 can be used to determine a first-order approximation of the inductance [Johnson and Graham, 1993]. To determine a more accurate model, a three-dimensional simulator or measurements from the laboratory should be used.

$$L_{\text{short gap}} \approx 5D \ln \frac{D}{W} \qquad \text{nanohenries} \qquad (6.1)$$

where L is the inductance, D the current divergence length, and W the gap width. Assuming that the discontinuity appears in the middle of a long net, the degradation of an input step can be approximated using the equation

$$T_{10-90L/R} \approx 2.2 \left(\frac{L_{\text{short gap}}}{2Z_o} \right) \qquad (6.2)$$

The following equation can be used to estimate the rise or fall time after it has passed over the gap, where T_{input} is the rise or fall time (10 to 90%) of the signal before the gap and T_{gap} is the rise or fall time seen after the gap.

$$T_{\text{gap}} \approx \sqrt{T_{10-90L/R}^2 + T_{\text{input}}^2} \qquad (6.3)$$

If the gap is electrically close to a receiver capacitance, three time constants ($\sqrt{LC}$) should be used to estimate the 10 to 90% edge rate in place of equation (6.2).

It should be noted that routing more than one line over a gap in the reference plane should never be done on high-speed signals, and the reference plane under the signals should remain continuous whenever possible. The reason the short gap equations were presented is because it is sometimes necessary to route over package degassing holes or via antipads in some designs. The formulas above can help the designer estimate the effects. If traversing a ground gap is unavoidable, the effect can be minimized by placing decoupling capacitors on both sides of the signal line to provide an ac short across the gap. This will effectively shorten the gap length (D) by providing an ac short across the gap. However, it is usually impossible to place a capacitor between each line on a bus.

Examination of a signal line traversing a gap in the ground plane leads to some general conclusions about nonideal return paths.

RULES OF THUMB: Nonideal Return Paths

- A nonideal return path will appear as an inductive discontinuity.
- A nonideal return path will slow the edge rate by filtering out high-frequency components.
- If the current divergence path is long enough, a nonideal return path will cause signal integrity problems at the receiver.
- Nonideal return paths will increase current loop area and subsequently exacerbate EMI.
- Nonideal return paths may significantly increase the coupling coefficient between signals.

6.1.3. Signals That Change Reference Planes

Another nonideal return path that is very often overlooked is a signal changing reference planes. For example, Figure 6.5 shows a CMOS output driving a transmission line. The transmission line is initially referenced to the power plane (V_{dd}). The transient return current, I_m, is induced on the power reference plane and flows directly beneath the signal line for a distance of X. The return current is shown as a negative value flowing in the same direction as the signal current. As the signal transitions to a ground plane reference, the return current cannot flow directly beneath the signal line. Subsequently, just as with the reference plane gap shown in Figure 6.2, the return current will flow through the path of least inductance so that it can complete the loop. In this case, however, instead of flowing around a gap in the

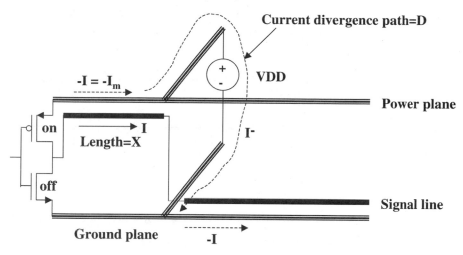

FIGURE 6.5 Current paths when a signal changes a reference plane.

reference plane, it will find the easiest path between planes. In this particular case, the easiest path is the V_{dd} power supply. It is easy to draw parallels between a signal switching reference planes and a signal traversing a gap. Obviously, since the current loop is increased, excess series inductance will be induced. This will degrade the edges and will cause a signal integrity problem in the signal waveform if the return path is long enough. Additionally, this particular example assumes that the return path is through an ideal voltage supply. In reality, the return path will be inductive and will probably be through a decoupling capacitor, not the power supply. Changing reference planes should always be avoided for high-speed signals. If it is absolutely necessary, however, the decoupling capacitance should be maximized in the area near the layer change. This will minimize the return current divergence path.

6.1.4. Signals Referenced to a Power or a Ground Plane

This type of return path discontinuity is almost never accounted for in design and can lead to severe problems. Initially, let's examine what happens to the return currents when a CMOS output is driving a ground-referenced microstrip transmission line as depicted in Figure 6.6. If the system is at steady state low and switches high, the return current in the ground plane must flow through the capacitor to complete the loop. The capacitor represents the local decoupling capacitance, which will usually be at or near the die. If this capacitance is not large enough, or there is significant inductance in the path, or if the closest decoupling capacitor is far away, the signal integrity will be affected adversely.

When designing a system, the engineer should examine the return currents very carefully and ideally should choose a stackup that will minimize transient

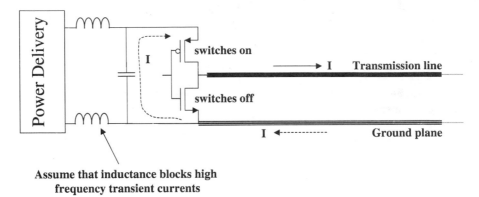

FIGURE 6.6 Return current for a CMOS buffer driving a ground-referenced microstrip line.

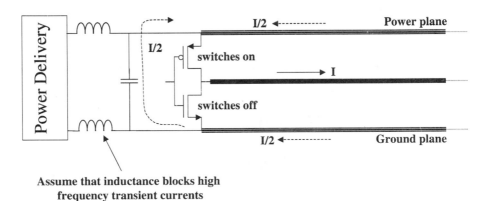

FIGURE 6.7 Return current for a CMOS buffer driving a dual-referenced (ground and power) stripline.

return current through the decoupling capacitor. For example, since a typical CMOS output buffer pulls an equal amount of current from power and ground, the optimal solution is to route the signal in a symmetric stripline that is referenced to both power and ground. If such a circuit is referenced to only a power or a ground plane, the return current through the decoupling capacitance will be equal to the mirror current in the reference plane as shown in Figure 6.6. If the line is symmetrically referenced to power and ground, as shown in Figure 6.7, the maximum current flowing through the decoupling will be approximately $\frac{1}{2}I$.

Figure 6.8 shows a GTL buffer driving a microstrip line that is referenced to the ground plane. The capacitance C represents the local decoupling capacitance at the I/O cell of the die. The network to the left of the driver represents a simplified power delivery network that is discussed in Section

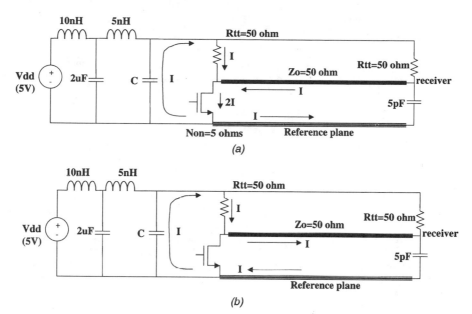

FIGURE 6.8 Return current for a GTL buffer driving a ground-referenced microstrip line: (a) pull-down (NMOS switched on); (b) pull-up (NMOS switched off).

6.2. In this example it is assumed that the inductance of the power supply blocks the high-frequency ac signals, which forces the return current to flow through the capacitance. The figure of merit is the amount of current forced through the decoupling network. It is optimal to choose the reference plane so that the minimum amount of current is forced through the decoupling capacitors.

Example 6.1: Calculation of Return Current for a GTL Bus High-to-Low Transition. Refer to Figure 6.8. Assume that impedance of the NMOS can be approximated by a 5-Ω resistor and the bus is just switching from steady state high to low. When the circuit pulls the net low, the voltage across the NMOS can easily be calculated.

$$V_{Low} = V_{dd} \frac{R_{NMOS}}{[R_{tt} \cdot Z_0/(R_{tt} + Z_0)] + R_{NMOS}} = (5)\frac{5}{25 + 5} = 0.833 \text{ V}$$

$$I_{NMOS} = \frac{V_{Low}}{R_{NMOS}} = \frac{0.833}{5} = 166 \text{ mA}$$

$$I_{R_{tt}} = \frac{V_{dd} - V_{Low}}{R_{tt}} = 83 \text{ mA}$$

Therefore, for the pull-down condition in Figure 6.8, the current forced though the decoupling capacitance is 83 mA. This is deduced by summing currents

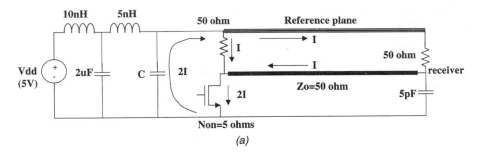

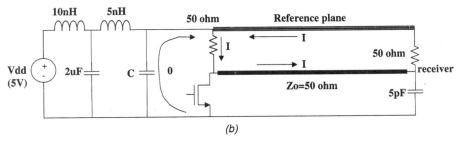

FIGURE 6.9   Return current for a GTL buffer driving a power referenced microstrip line: (*a*) pull-down; (*b*) pull-up.

at the NMOS-transmission line junction and at the NMOS-ground plane junction.

Figure 6.9 shows the return currents for a power-plane-referenced microstrip line. Notice that no current is forced through the decoupling capacitance for the pull-up transition, but twice the current is forced through the decoupling capacitor for the pull-down transition. Subsequently, it can easily be concluded that it is preferable to route a GTL bus over ground planes unless it can be proven that the local decoupling is adequate to pass the return currents without distorting the signal.

Figure 6.10 are some simulations showing the difference between ground- and power-referenced buses for the circuits shown in Figures 6.8 and 6.9. Notice that the ground-plane-referenced microstrip exhibits much cleaner signal integrity than the power-plane-referenced signal. Also note that the signal integrity for both cases improves significantly when the local decoupling capacitance C is increased so that the capacitor looks like a short to the transient return currents.

Since it is important to model all current return paths explicitly through the power delivery system, it is necessary to use only one ground in the system simulation and explicitly force all current to flow as they would in the real system. There are two ways to do this. The easiest way is to construct a traditional *LRC* model as depicted in Figure 6.11*a* and tie all the grounds to a common

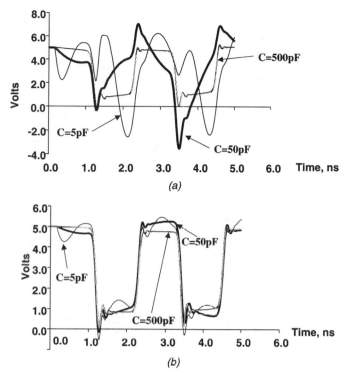

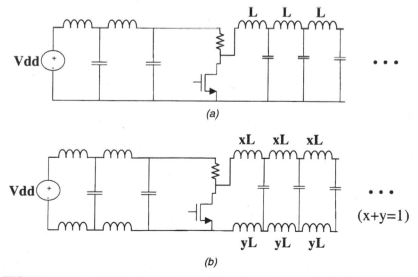

FIGURE 6.10 Signal integrity as a function of local decoupling capacitance: (*a*) power-plane-referenced microstrip; (*b*) ground-plane-referenced microstrip. (Circuits simulated are shown in Figures 6.8 and 6.9.)

FIGURE 6.11 (*a*) Unbalanced versus (*b*) balanced transmission line models.

point so that the current is forced to flow through the power delivery system. If a specialized transmission line model included in a simulator is used, however, the ground-referenced points cannot necessarily be tied to anything other than perfect ground, because it might cause convergence problems, depending on the model. The alternative, which is technically more correct, is to use a balanced transmission line model instead of the traditional unbalanced model. A balanced model has inductance in the top and bottom portions so that an explicit ground return path can be modeled. Unfortunately, this typically takes a very long time to converge during simulation. Convergence can be expedited, however, if resistive elements are included in the simulation. Figure 6.11 shows the difference between a balanced and an unbalanced return path model. In the balanced model, the inductance (L_{11}) is split up between the signal layer and the ground layer. The amount of inductance in each layer depends on the current distribution. Since this is difficult to calculate without a three-dimensional simulator, it is usually a valid approximation to include 80% of the inductance in the signal layer and 20% of the inductance in the plane layer. In most designs, simulations are performed with the unbalanced topology.

It should be noted that accurate modeling of this return path phenomena depends on the power delivery network. Subsequently, it is prudent to read the subject of power delivery (Section 6.2) prior to creating any models.

6.1.5. Other Nonideal Return Path Scenarios

There are significantly more nonideal return path scenarios than can be covered in this book. However, the discussion so far should have provided a basic understanding of the effects. Here is a brief list of some return path discontinuities not discussed.

- Meshed power/ground planes
- Signals passing over or near via antipads
- Transitions from a dual-referenced (power and ground) stripline to a single-referenced (just power or ground) microstrip or offset stripline
- Transitions from a package to a board

6.1.6. Differential Signals

When a return path discontinuity exists, the effect can be minimized through the use of differential signaling (odd-mode propagation). To explain why, refer to the odd-mode electric field shown in Figure 3.10. Notice that the electric fields always intersect the ground planes at 90° angles. Also notice that it is possible to draw a line through the electric field patterns between the conductors such that the electric field lines are always perpendicular to it (only in odd mode). This illustrates the fact that a virtual ground exists between the two conductors in an odd-mode pattern. When an odd-mode signal is propagating over a nonideal return path, the virtual ground provides an alternative reference to the signals so that a return path that is not perfect affects the signal quality to a lesser degree. Furthermore, the current through the decoupling capacitance will be minimized because when one driver is pulling current out of a supply, the complimentary driver is dumping current into the same plane. Subsequently, the currents through the decoupling capacitors will cancel. Differential signaling will dramatically improve the signal quality when nonideal effects from return paths, connectors or packages are present. Unfortunately, practical limitations such as package pin count, board real estate and receiver circuit complexity often negate the advantages gained from differential signaling.

6.2. LOCAL POWER DELIVERY NETWORKS

Power delivery will initially be discussed only as it pertains to the performance of the I/O buffers and the interconnects on the bus. The reader should note that although the analysis of local power delivery networks share much in common with system-level power delivery, the focus is different. The focus is supplying the required high-frequency current to the I/O buffers.

Consider the *LC* ladder to the left of the GTL output buffer depicted in Figure 6.8. This is a very simplified network intended to represent the power supply as seen through the eyes of the output buffer. The ideal voltage supply represents the voltage at the system voltage regulator module (VRM), which provides a stable dc voltage to the system. The first 10-nH inductor crudely represents the inductive path between the VRM and the first bulk decoupling capacitor, whose value is 2 μF. The 5-nH inductor represents the path inductance between the bulk decoupling capacitor and the I/O cell on the die. As mentioned above, the capacitor *C* represents the local decoupling on the die at the I/O cell. As mentioned earlier, this is only a very crude representation of a power delivery system; however, it is adequate to achieve first-order approximations and is excellent for instructional purposes. A full model would require extraction from a three-dimensional simulator (or measurements) and would be implemented in a simulator with a mesh of inductors that resembles a bedspring. Alternatively, full-wave simulators that use the finite-difference time domain (FDTD)–based algorithms can be used to optimize the local power delivery system. The FDTD simulator approach will yield more accurate responses of the board and package interactions; however, it is usually very difficult or impossible to model the active circuitry correctly in such simulators. To approximate the first-order effects, however, it is adequate to represent the power delivery system in the simplified manner discussed here and to ensure that the decoupling capacitance is adequate so that the local power delivery system will not affect the signal integrity.

The problem is that when the output buffer switches, it quickly tries to draw current from the VRM. The series inductance to the VRM will limit the current that can be supplied during the switching time. If the inductance is large enough, it will essentially isolate the output buffer from the power supply when current is drawn at a fast rate. If the inductance is high enough, it essentially looks like an open circuit during fast transitions and will block the flow of current. Subsequently, the power seen at the I/O cell will droop because the local power delivery system cannot supply the required current. This effect has many names, including *power droop*, *ground bounce*, and *rail collapse*. Whatever it may be called, it can devastate the signal integrity if not accounted for properly.

Ideally, the elimination, or the significant reduction of the series inductance, will solve the problem of rail collapse. However, reality eliminates the ideal solution because it is impossible to place a VRM in close proximity to every I/O cell. The next best thing is to place decoupling capacitors as close as possible to the component. The decoupling capacitors will be charged up by the VRM and will act like local batteries or mini power supplies to the I/O cell. If the decoupling capacitance is large enough and the series inductance to the decoupling capacitors is small, they will supply the necessary current during the transition and preserve the signal integrity. Figure 6.10 depicts the signal integrity as a function of local I/O capacitance. The capacitance C represents the local capacitance at the I/O cell as depicted in Figures 6.8 and 6.9. Notice that the signal integrity gets progressively better as the local capacitance is increased.

Another effect of power droop is a timing push-out in the form of a ledge, as shown in Figure 6.12. This effect is quite common, especially in CMOS

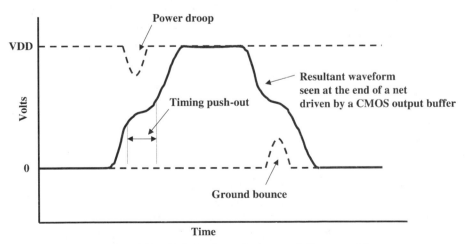

FIGURE 6.12   Telltale sign of a power delivery problem.

output drivers. As the CMOS gate pulls up, it will pull current out of the power supply at a fast rate. If the series inductance to the supply is large enough, it will limit the current and cause a decrease in the voltage seen at the drain of the PMOS. Subsequently, the voltage at the output will decrease, causing a ledge in the rising edge of the waveform. Similar effects occur when the NMOS is pulling the net to ground and attempting to sink current. A waveform of this shape (Figure 6.12) seen in the laboratory is a telltale sign that the decoupling of the power supply is inadequate. Be certain, however, that the waveform is on the receiving end of a net and not at the driver. A similar waveform is expected at the driver of a transmission line due to the initial voltage divider between the output impedance and the transmission line impedance (e.g., see Figure 2.14). A ledge can also be caused by other effects such as a long stub (e.g., Figure 5.15). Do not get these phenomena confused.

Fundamentally, this leaves us with two critical components of the local power delivery system as it pertains to the output buffers and signal integrity: the value of the nearest decoupling capacitance and the inductive path to that capacitance. Often, the major decoupling mechanism, depicted in Figures 6.8 through 6.10, is the natural on-die capacitance. Other times, it is the closest decoupling capacitors on the board. At any rate, it is essential to place the capacitance as close as possible to the component. Sometimes this is achieved by designing a chip package with an area underneath that is devoid of pins for the placement of decoupling capacitors directly beneath the die. These are sometimes referred to as *land-side capacitors* and can dramatically improve the local power delivery system.

Another consideration when designing a system is power delivery resonance. This occurs when the decoupling capacitance and the series inductance resonates like a simple tank circuit. (More information on decoupling resonance is provided in Section 10.3.1.) This will be as shown in Figure 6.10 when the capacitance at the die is 5 pF. It is much more evident in the power-plane-referenced example; however, it is definitely noticeable in the ground-plane-referenced circuit. The resonance occurs between the 5-pF capacitance at the die and the 5-nH series inductance to the first bulk decoupling capacitance. This resonance of a parallel L and C forms a high impedance pole at which it will be very difficult to draw power. In this case the resonance is calculated as

$$f_0 = \frac{1}{2\pi\sqrt{LC}} = \frac{1}{2\pi\sqrt{5 \text{ pF} \cdot 5 \text{ nH}}} = 1.0 \text{ GHz} \qquad (6.4)$$

It should be noted that the resonance might be calculated differently in other systems. Only careful examination of the power delivery system will yield the proper resonance equation. It is vital that the resonance of the power supply network be considered during design so that it will not adversely affect signal integrity.

6.2.1. Determining the Local Decoupling Requirements for High-Speed I/O

Ideally, it is optimal to design enough on-die capacitance into the silicon to provide adequate power delivery to the output buffer. However, it is not always possible to do so. When the on-die capacitance is not sufficient, it is necessary to compensate with external capacitors. The minimum on-die capacitance that is required to eliminate signal integrity problems can be approximated using the following process.

1. Determine the maximum amount of noise your logic can tolerate due to local power supply fluctuations. Remember, power droop is only one component of the total noise, so de-rate this number accordingly. We will call this number ΔV.

2. Determine the maximum amount of return current that will be forced to flow through the decoupling capacitance. We will call this number ΔI.

3. Subsequently, $\Delta V / \Delta I$ is the maximum impedance that the path through the local on die decoupling can tolerate.

4. Approximate the transient frequency the on-die capacitance must pass. There are many ways to do this. One method is to use the equation

$$F_{\text{tran}} \approx \frac{0.35}{T_{\text{slow}}} \qquad (6.5)$$

which is derived in Appendix C, where T_{slow} is the *slowest* edge rate the driver will produce. It is important to choose the slowest edge rate because a capacitor will pass high frequencies. If the fastest edge rate is used in this calculation, signal integrity problems may arise for components with slower edge rates because the capacitor will not pass the lower-frequency components of the slower edge rate. Alternatively, the engineer may choose use the fundamental maximum bus frequency or Fourier analysis methods to determine the frequencies that must be decoupled by the capacitor.

5. The minimum on-die capacitance per I/O cell is calculated as

$$C_{\text{min}} \approx \frac{\Delta I}{2\pi F_{\text{tran}} \Delta V} \qquad (6.6)$$

If the on-die capacitance per I/O cell is smaller than C_{min}, the board or land-side decoupling capacitance must compensate. The main problem, as discussed previously, is the series inductance to the decoupling capacitor. This inductance must be kept small enough so that it won't limit the current to the nearest capacitor. Unfortunately, since it will be almost impossible to provide a board-level decoupling capacitor for each I/O cell, each capacitor must provide the decoupling for several output buffers. Furthermore, the inductive path to that capacitor is also shared. To determine the requirements of the closest board-level decoupling capacitors, both the inductive path and the capacitance value must be investigated. The inductance must be small enough and the capacitance must be large enough. The following procedure approximates the requirements assuming the on-die capacitance is not adequate.

1. Again, determine the maximum amount of noise your logic can tolerate due to local power supply fluctuations. We will call this number ΔV.

2. Determine the maximum amount of return current that will be forced to flow through the inductive path to the decoupling capacitance. One way to approximate this is to determine the total number of I/O cells per board-level decoupling capacitors. Since the return current of each I/O will be forced to flow through the capacitor, the sum will be ΔI_{sum}. For example, if the ratio of I/O cells to decoupling capacitors is $3:1$, the return current of three output buffers will be forced the flow through each capacitor.

3. Determine the transient frequency of the edge as in step 4 of the earlier list in this section. It is important to note, however, that the *fastest* rise/fall time should be used because an inductor will pass low frequencies.

4. Calculate the maximum tolerable series inductance to the cap as

$$L_{max} \approx \frac{\Delta V}{2\pi F_{tran}\Delta I_{sum}} \tag{6.7}$$

5. Calculate the minimum value of the decoupling capacitance with equation (6.6) using the value ΔI_{sum} in place of ΔI. Remember, for the minimum capacitance value, use the *slowest* rise or fall time. This will ensure the capacitance is large enough to pass all frequencies.

The value of L_{max} will obviously depend on the distance between the I/O cells and the decoupling capacitors. It will also depend on the package, socket, via, and capacitor lead inductance. The package, via, and socket inductance can be approximated using the techniques discussed in Chapter 5. The series inductance of the capacitor itself can either be measured or provided by the manufacturer. The determination of the plane inductance will require a three-dimensional simulator or a measurement to determine accurately; however, it can be approximated to the first order by using either a two-dimensional simulator or plane inductance equations. It should be noted that the total inductance contributed by sockets, chip packages, bond wires, vias, and so on, is typically much higher than the plane inductance.

To crudely approximate the plane inductance between the package power pin and the nearest decoupling capacitor, the following steps can be followed.

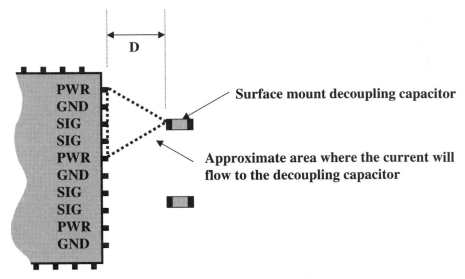

FIGURE 6.13   Estimating the area where the current will flow from the component to the decoupling capacitor.

1. Determine the ratio of local decoupling capacitors to output buffers. This will provide an estimate of the number of signals decoupled by each capacitor.

2. Approximate the area on the PCB power plane where the current will flow. This may prove to be very difficult without a three-dimensional simulator. The area is important because it will be used to calculate the plane inductance of the path between the power pin and the decoupling capacitor. If sound engineering judgment is used, the area may be approximated within an order of magnitude. Since the current will always travel through the path of least inductance, a straight line can usually be assumed. The problem is estimating how much the current will spread out. One way of roughly estimating the spread is shown in Figure 6.13. In Figure 6.13, the ratio of power pins to capacitors is 2 : 1. Subsequently, the approximate area where the current will flow can be approximated by the triangle shown.

3. Determine the plane inductance between the package power pin and the nearest decoupling capacitor. This can be approximated by constructing a rectangle with the same area as the triangle shown in Figure 6.13 and simulating it in a two-dimensional solver as a wide transmission line with a length of D. Alternatively, the following equation can be used [Johnson and Graham, 1993]:

$$L_{\text{plane segment}} \approx 31.9 \frac{DH}{W} \qquad \text{nanohenries} \qquad (6.8)$$

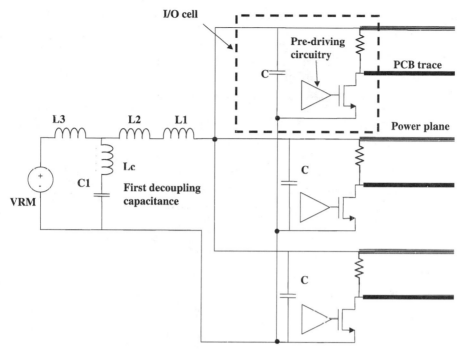

FIGURE 6.14 Equivalent circuit of three GTL drivers and a component-level power delivery system. C, on-die I/O cell capacitance; C_1, first-level decoupling capacitor; L_1, package and socket equivalent inductance; L_2, PCB plane inductance from package power pin to decoupling capacitor; L_3, inductance to the VRM; L_c, lead inductance of the capacitor.

where D is the length of the plane segment in inches, H the separation between the power and ground planes (in inches), and W the width in inches. This should provide an estimate that is accurate to at least one order of magnitude. If a more accurate number is required, a three-dimensional simulator or laboratory measurements are required.

Figure 6.14 shows an equivalent circuit of a GTL bus where three output buffers are sharing one board level decoupling capacitor. A full model would include the crosstalk in the package and socket.

6.2.2. System-Level Power Delivery

The subject of system-level power delivery is discussed only briefly in this book. The subject is also discussed in Chapter 10. Typically for interconnect design, the power delivery required for analysis of the bus performance is at the component level, as discussed in Section 6.2.1. This discussion is intended only to provide the designer with a fundamental understanding of system-level power delivery.

A system-level power delivery network is designed to provide stable and uniform voltages for all devices. It is very important that the system power level be kept uniform across the board because fluctuations in reference and/or supply voltages will significantly affect the timing and signal integrity of individual components. Such variations can lead to component-to-component skews that can consume precious timing budgets. These devices usually require a low-impedance voltage source that is able to supply the necessary current when the logic gates switch. If the components were connected directly to the power supply or the VRM, there would be no need to worry about systemwide power delivery. Unfortunately, the power supply of the system is distributed via a network of inductive planes and wires. Subsequently, when numerous gates switch simultaneously on several different components, the inductance and resistance of the planes and wires will exhibit high-impedance characteristics and may limit the amount of current that is available instantaneously. The purpose of system-level power delivery is to provide and adequate supply to satisfy all current demand of the system-level components.

There are essentially two levels of power delivery that must be accounted for in a digital system, the low frequency and the high frequency. The high frequency is required at the component level, which was described previously in this chapter with a focus on the I/O circuitry. However, the core logic of a device also requires high-frequency power delivery. Basically, the high-frequency component of the power delivery system must supply the instantaneous current demanded at the part. As described above, this is achieved by placing capacitors as close as possible to the component and maximizing the on-die capacitance. Subsequently, the capacitors will act like local batteries to provide the necessary current. These capacitors need to be large enough so that an adequate amount of charge can be stored in order to supply the required current. The problem is that once the capacitors are depleted of charge, they can no longer provide the current until they are charged up again. If there is a large inductive path between the high-frequency capacitors at the component and the voltage supply, they will not be charged up in time to provide current when it is required. The solution is to add more decoupling capacitance between the voltage supply (VRM) and the component. If done correctly, this will supply a charge reservoir that will recharge the component-level capacitors. This second tier of decoupling capacitance will usually be much larger than the component or first-tier capacitors. The bandwidth requirements of the decoupling network will decrease as the capacitors move away from the component. They don't need to support the full di/dt value of the logic; they only have to support a bandwidth high enough to recharge the first-tier capacitors before they are required to supply current to the local component. Sometimes it is required to have several tiers of decoupling capacitors between a system power supply and the components to guarantee a stable and uniform voltage reference across the board. As the capacitors move away from the component, the frequencies they must support become lower.

6.2.3. Choosing a Decoupling Capacitor

Capacitors are not perfect. Every discrete capacitor has a finite lead inductance associated with it that will cause the impedance of the decoupling capacitor to increase when the frequency gets high enough. They also have a finite resistive loss that will decrease the effectiveness of the capacitor [called effective series resistance (ESR)]. Furthermore, every capacitor is temperature sensitive, which may cause its dielectric properties to vary and cause wide swings in capacitance. The capacitance of a capacitor may also change slowly over time due to aging of the dielectric material. Finally, they will explode if subjected to excessive voltage. When choosing a bypass capacitor, it is important to comprehend each of these imperfections. The effect of the inductance and ESR can be calculated. Only the manufacturer can provide detailed information on temperature sensitivity, aging, and maximum voltage.

The figure of merit for a decoupling capacitor is its equivalent ac impedance. This is estimated using a root-mean-square approach including the impedance of the resistance, inductance, and capacitance:

$$X_{ac}(F) \approx \sqrt{R_{ESR}^2 + \left(2\pi F L - \frac{1}{2\pi F C}\right)^2} \qquad \text{ohms} \qquad (6.9)$$

where R_{ESR} is the series resistance of the capacitor, X_{ac} the equivalent ac impedance of the capacitor, and L the sum of the lead, package, and connector inductance of the discrete capacitor.

Figure 6.15 is a frequency response of a realistic capacitor. The impedance of the inductance, capacitance, and resistance is plotted versus frequency on a log-log scale so that it can be analyzed easily. The results of equation (6.9) are also plotted. Notice the bandpass characteristics. At low frequencies, the capacitor behaves like a capacitor. When the frequency gets high enough, the inductive portion takes over and the impedance increases with frequency.

The impedance of the capacitor will depend largely on the frequency content of the digital signal. Therefore, it is important to choose this frequency correctly. Unfortunately, it is not very straightforward in a digital system because the signals contain multiple frequency components. There are several ways to choose the maximum frequency that the bypass capacitor must pass. Some engineers simply choose the maximum frequency to be the fifth harmonic of the fundamental. For example, if a bus is running at a speed of 500 MHz, its fifth harmonic is 2500 MHz (see Appendix C). Usually, the rising and falling edges of a digital signal produce the highest-frequency component. Subsequently, the following equation (also explained in Appendix C) can be used to approximate the highest frequency that the bypass capacitor

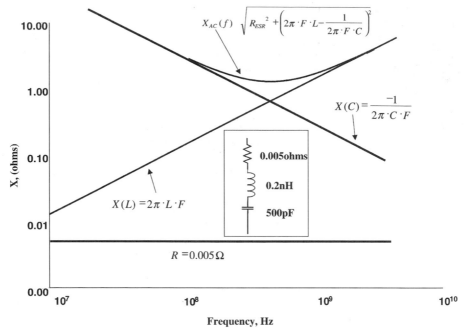

FIGURE 6.15   Impedance versus frequency for a discrete bypass capacitor.

must pass:

$$F_{\text{bypass}} = \frac{0.35}{T_{\text{rise/fall}}} \tag{6.10}$$

If the lead inductance or the ESR of the capacitor is too high, either choose a different one or place several capacitors in parallel to lower the effective inductance and resistance.

6.2.4. Frequency Response of a Power Delivery System

Of course, the frequency response of the entire power delivery network is what is important. The system power supply must look low impedance at all frequencies of interest. Figure 6.16 shows how the ac impedance will vary as a function of frequency for a simple power delivery system. Bode plot techniques were used to estimate the impedance as a function of frequency for the power delivery system. Note that the power delivery inductance, L_{pwr}, plays a significant role in the frequency response. In this analysis, the effect of the series resistance (ESR) was ignored. Another effect that is ignored in this analysis is the possibility of resonant poles that can significantly increase the impedance of the power delivery system. These resonant poles are covered briefly in Section 10.3.1.

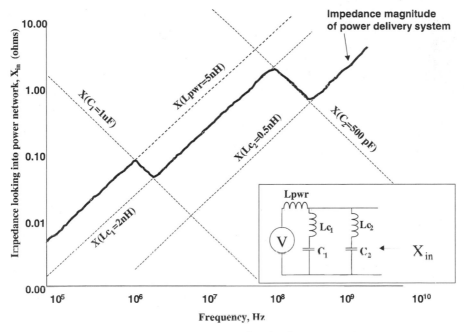

FIGURE 6.16 Frequency response of a simple power delivery system.

6.3. SSO/SSN

Simultaneous switching output noise (SSO), which is sometimes referred to as *simultaneous switching noise* (SSN) or *delta-I noise,* is inductive noise caused by several outputs switching at the same time. For example, a signal switching by itself may have perfect signal integrity. However, when all the signals in a bus are switching simultaneously, noise generated from the other signals can corrupt the signal quality of the target net.

SSN is typically very difficult to quantify because it depends heavily on the physical geometry of the system. The basic mechanism, however, is the familiar equation

$$V_{SSN} = N L_{tot} \frac{dI}{dt} \tag{6.11}$$

where V_{SSN} is the simultaneous switching noise, N the number of drivers switching, L_{tot} the equivalent inductance in which current must pass, and I the current per driver. When a large number of signals switch at the same time, the power supply must deliver enough current to satisfy the sudden demand. Since the current must pass through an inductance, L_{tot}, a noise of V_{SSN} will be introduced onto the power supply, which in turn will manifest itself at the driver output.

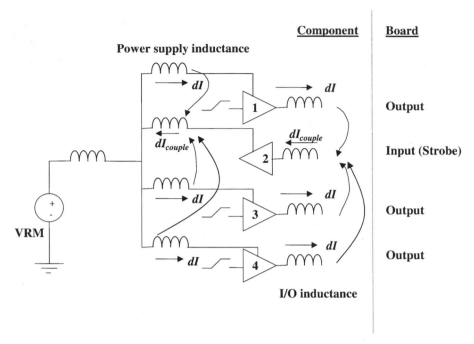

FIGURE 6.17   Simultaneous switching noise mechanisms.

SSN can occur at both the chip level and the board level. At the chip level, the power supply is not perfect. Any sudden demand for current must be supplied by the board-level power though the inductive chip package and lead frame (or whatever the connecting mechanism happens to be). On the board level, sudden current demands must be supplied through inductive connectors. As discussed in Chapter 5, any current flowing through a connector must be supplied and return through the power and ground pins, which will induce noise into the system. Furthermore, as discussed in Section 6.1, a nonideal return path will also cause an effective increase in the series inductance in the vicinity of the discontinuity. Furthermore, if the return path discontinuity forces the return current from several outputs to flow through a small area, SSN will be even more exacerbated.

Many of the mechanisms that contribute to SSN were discussed in Sections 6.1, 6.2, and 5.2.4. The area that needs further explanation is the chip/package-level SSN. To explain how SSN will affect signals on the chip, refer to Figure 6.17. The inductors represent the equivalent inductance seen at the power connections and the output of each I/O. As drivers 1, 3, and 4 switch simultaneously, several inductive noises will be generated. First, as the transient current flows through the power supply inductors, di/dt noise will be generated. This noise will be coupled to the power connections of quiet nets. It is possible for the power supply to get so noisy that it causes core logic to flip

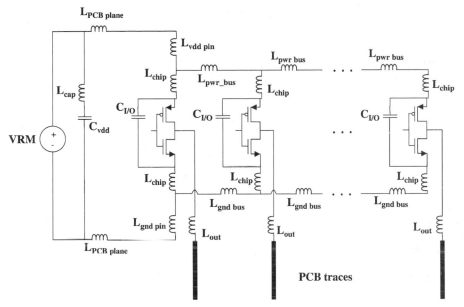

FIGURE 6.18 Model used to evaluate component level SSN/SSO for a CMOS-driven bus.

state. In another failing case, the input of a strobe or a clock receives enough noise that exceeds its threshold voltage, causing a false trigger. SSN will also distort the signal integrity, which can cause gate delays, which will sometimes be manifested as shown in Figure 6.12.

SSN can be a very elusive noise to characterize. There are not many methods for quick approximations to get an easy assessment of SSN. Only careful examination of your package and power delivery system and detailed simulations can lead to a reasonable assessment of the magnitude of SSN. Even when attempts are made to characterize the noise accurately, it is almost impossible to determine an exact answer because the variables are so numerous and the geometries that must be assessed are three-dimensional in nature and depend heavily on the individual chip package (or connector) and the pin-out. Because of the difficulty of this problem, it is recommended that SSN be evaluated using both simulation and measurements. Subsequently, only general rules can be used to control this noise source.

Figure 6.18 is a generic model that can be used to evaluate SSN in a CMOS bus. The capacitors, $C_{I/O}$, are the inherent on-die capacitance for each I/O cell. L_{chip} represents any inductance seen on the chip between the CMOS gate and the power bus. $L_{pwr\,bus}$ represents the inductance of the power distribution on the die and package. $L_{gnd\,bus}$ represents the inductance of the ground distribution on the die and on the package. $L_{V_{dd}\,pin}$ and $L_{gnd\,pin}$ represent the inductance of the power and ground pins on the package. $L_{PCB\,plane}$ represents the inductive path between the pin and the nearest decoupling capacitor. L_{cap} represents the series inductance of the decoupling capacitors, and $C_{V_{dd}}$ represents the board-level decoupling capacitors. Finally, L_{out} represents the series inductance of the package seen at the I/O outputs. It should be noted that all mutual inductance values should be included in this model. Furthermore, the number of gates simulated should be equal to the number of gates that share the same power and ground pins.

Connector SSN is simulated as described in Section 5.2.4. It is important, however, that the decoupling capacitors and the inductive path to the capacitors be accounted for properly. If the general guidelines for connector design described in Section 5.2.6 are adhered to, connector-related SSN will be minimized. Nonideal return paths cause board-level SSN, which effectively increases the series inductance of the net. Subsequently, the avoidance of any nonideal return paths is critical. If a nonideal return path is unavoidable, the board must be heavily decoupled in the area of the discontinuity.

6.3.1. Minimizing SSN

The following steps may be taken to decrease the effects of SSN:

1. If possible, use differential output drivers and receivers for critical signals such as strobes and clocks. A differential output consists of a pair of signals that are always switching opposite in phase (odd mode). A differential receiver is simply a circuit that triggers at the crossing of two signals. This will eliminate common-mode noise and increase the signal quality dramatically. Differentially routed transmission lines will also be more immune to coupled noise and nonideal return paths because the odd mode sets up a virtual ground between the signals. It will also help even out the current drawn from the power supply.

2. Maximize the on-die capacitance. This will provide a charge reservoir that is not isolated by inductance. If the on-die capacitance is large enough, it will act like a local battery and compensate for a sudden demand in current.

3. Maximize the decoupling local to the component. Use land-side or die-side capacitors, if possible. Place the board capacitors as close as possible to the power and ground pins.

4. Assign I/O pins so there is minimum local bunching of output pins. Choose a pin-out that maximizes the coupling between signal and power/ground pins. Maximize the number of power and ground pins. Place power and ground pins adjacent to each other. Since current flows in opposite directions in power and ground pins, the total inductance will be reduced by the mutual inductance.

5. Reduce the edge rates. Be careful, though; this can be a double-edged sword. Slower edges are more susceptible to core noise. Laboratory

measurements have shown that core noise can couple into the predriver circuitry during a rising or falling transition and cause jitter. The slower the transition, the more chance noise will be coupled onto the edge.

6. Provide separate power supplies for the core logic of a processor and the I/O. This will minimize the chance of SSN coupling into the core (or vice versa) and causing latches to switch state falsely.

7. Reduce the inductance as much as possible. Using fat power buses and short bond wires can do this.

8. When designing a connector, follow the guidelines in Section 5.2.6.

9. Minimize the inductance of the decoupling capacitors.

10. Avoid nonideal return paths like the plague.

11. Reference the PCB traces to the appropriate planes to minimize the current forced to flow through the decoupling capacitors, as in Section 6.1.4.

CHAPTER 7

Buffer Modeling

A large fraction of this book has explained how to model and understand various elements of a digital system. In this section we explain how to model the driving circuits and incorporate the models into the system interconnect simulations. Figure 7.1 shows one example of a buffer tied to a transmission line through a series resistor R_s. In this example, the series resistor R_s plus the impedance of the output buffer Z_s should roughly equal the characteristic impedance of the transmission line Z_o in order to have matched termination. Accounting for the behavior of various buffer types is the subject of this chapter.

Conceptually, drivers and receivers have largely been treated as fixed-impedance devices up to this point. Although this is useful for deriving the many concepts contained elsewhere in the book, the approximation ignores the fact that the buffer impedance values are dynamic and depend on many variables. In this chapter we pick a few buffer types and show how models are derived that can be incorporated into system-level simulations. These concepts can be extended to buffer types not presented here. We do not provide details on actual implementation of models. Instead, we explain some basic concepts, to allow the reader to understand the models and how the simulators may interpret them. This will allow the reader to merge his or her understanding with the function of the simulator.

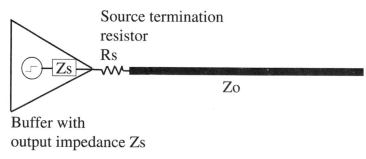

FIGURE 7.1 Generic buffer implementation.

7.1. TYPES OF MODELS

There are three general types of models for buffers used for simulation in digital systems: (1) linear model, (2) behavioral model, and (3) full transistor model. In this chapter we focus on linear and behavioral models. Full transistor models generally include detailed descriptions of transistors, and are beyond the scope of this book. Several references on transistor details and transistor models are available. The linear model and behavioral model are similar in that they are constructed from curves describing the voltage, current, and time behavior. These models generally can be constructed without proprietary information and are often preferable for this reason. Vendors can distribute buffer models without giving away sensitive information. Also, linear and behavioral models tend to be compatible over a wide variety of simulators, which allows tool independence. The models also run much faster than do full transistor models.

Full transistor models are generally created for some sort of SPICE-like simulator and contain detailed information about the transistors. These models are generally the most accurate, but simulation times and complexity often prohibit their use. Also, the information in the models is not in a format that is readily extractable for understanding the buffer characteristics. Generally, if buffers are completely undefined at the beginning of a design cycle, linear models should be created and simulations performed to determine the desired buffer output impedance and edge rates. The linear buffer model is only valid, however, if a buffer with controlled and roughly linear impedance is desired for the final design. If control of impedance is not important, use of a linear model may not be merited. The information gained from simulations using linear buffer models can later be communicated to buffer designers and used to design actual circuits. More detailed information from either full transistor simulation or measurements can later be used to create more detailed behavioral models. If, however, the buffer to be used in a design already exists, initial simulations should be performed using a behavioral model or a full transistor model.

7.2. Basic CMOS Output Buffer

The CMOS output buffer is very common in digital designs. Subsequently, it is the buffer chosen to demonstrate the basic principles of buffer modeling.

7.2.1. Basic Operation

The basic CMOS output buffer is shown in Figure 7.2. When the input is high, the output is driven low through the N-type device. When the input is low, the output is driven high through the P-type device. Conceptually, driving devices have been implicitly modeled elsewhere in this book, as shown in Figure 7.3. The impedance of the CMOS buffer shown in Figure 7.2 is dependent

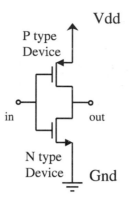

FIGURE 7.2   Basic CMOS output buffer.

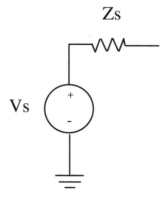

FIGURE 7.3   General method of describing buffers elsewhere in this book.

on the instantaneous operating conditions of the device. In other words, the impedance is dynamic. The instantaneous output impedance can be thought of as the slope (or the inverse slope) of the current–voltage (I–V) characteristic curve of the buffer. Some general curves for N and P devices are shown in Figure 7.4. The reader will recall the equations governing operation of the CMOS FET transistors as shown in equations (7.1) through (7.4) [Sedra and Smith 1991]. The two regions of operation of a FET are the triode region and the saturation region. An N device is in the triode region if $v_{ds} \leq v_{gs} - V_t$ and in the saturation region if $v_{ds} \geq v_{gs} - V_t$, where V_t is the threshold voltage of the device. A P device is in the triode region if $v_{ds} \geq v_{gs} - V_t$ and the saturation region if $v_{ds} \leq v_{gs} - V_t$.

For the triode region (see Figure 7.4),

$$i_d = K[2(v_{gs} - V_t)v_{ds} - v_{ds}^2] \tag{7.1}$$

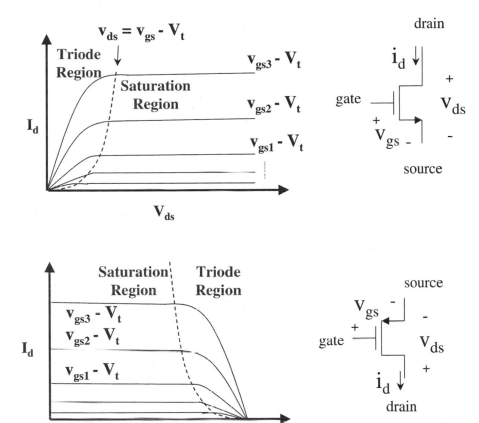

FIGURE 7.4 NMOS and PMOS *I–V* curves.

For CMOS devices, K is $\frac{1}{2}\mu C_{ox}(W/L)$, where μ is the mobility of charge carriers (i.e., electrons or holes) in the silicon, C_{ox} the oxide capacitance, and W and L the width and length of the transistor being considered. For the saturation region (see Figure 7.4),

$$i_d = K(v_{gs} - V_t)^2(1 + \lambda v_{ds}) \qquad (7.2)$$

where λ is a positive parameter that accounts for the nonzero slope (i.e., the output impedance) of the curve in the saturation region. The output impedance in the saturation region is defined as

$$\frac{1}{Z_{out}} = \frac{\partial i_d}{\partial v_{ds}} = \lambda K(v_{gs} - V_t)^2 \qquad (7.3)$$

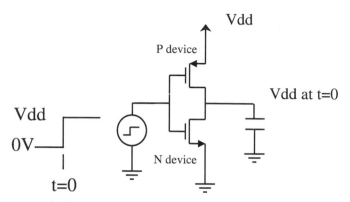

FIGURE 7.5   CMOS output buffer driving a load.

This is *not* the output impedance that should be considered in a model for the basic CMOS buffer, since the CMOS buffer is typically not operated in the saturation region. The output impedance in the saturation region is mentioned here only for completeness and later relevance. Z_{out} in the saturation region is typically very high, and thus $\lambda \simeq 0$ is usually assumed resulting in

$$i_d = K(v_{gs} - V_t)^2 \tag{7.4}$$

for the saturation region.

In order to derive models, let us briefly describe the operation of the basic CMOS buffer. Consider Figure 7.5, which shows the input to the buffer transitioning from low to high with an initial condition of V_{dd} volts at the output of the buffer. V_{dd} will be present at the gate of both the P and N devices; thus the N device will have a gate-to-source voltage, V_{gs}, of V_{dd}. The N device will thus be in an "on" condition and will begin to discharge the load to ground. Subsequently, this transition is often called the *pull-down*. The initial point ($t = 0$) on the transistor curve can be found on the $V_{gs} = V_{dd}$ curve at $V_{ds} = V_{dd}$, as shown in Figure 7.6a. At time $t = 0$, the device operating point will be as shown and the N device will conduct current to discharge the load. As the load is discharged, V_{ds} will decrease and the operating point will shift downward on the curve until the load is discharged to very close to zero volts. The curve for the P device is also shown with $V_{gs} = 0$ V. The current conducted by the P device at $V_{gs} = 0$ will be equal or very close to zero. The intersection of the P and N curve will be the final, steady-state value of the output when the input to the gate is V_{dd}.

Conversely, when the input to the buffer transitions to zero volts, the N device will turn off and the P device will turn on, which will charge the load to V_{dd}. This is known as the *pull-up* transition. The transistor curves shown in Figure 7.6b shows the steady-state output value of the buffer when the input is low. The impedance at any point along the discharge is the inverse

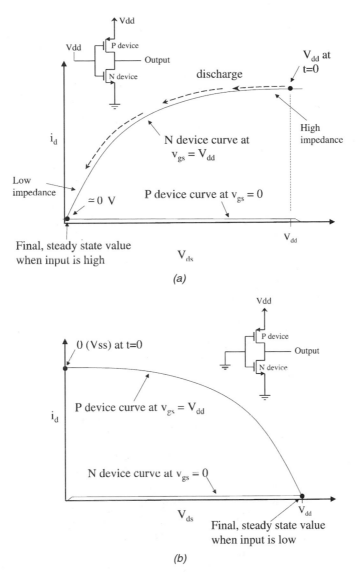

FIGURE 7.6 Operation of the CMOS output buffer when the input voltage is (*a*) high and (*b*) low.

of the slope of the *I–V* curve. In Figure 7.6*a*, the impedance varies widely, from a very high value at $t = 0$ to a lower value as the load discharges. This wide variation in impedance is, of course, not optimal if a controlled range of impedance is desired from the buffer for matching to a transmission line. Ideally, for precisely controlled impedance, the *I–V* curve will be that of a fixed resistor, which is a straight line, as shown in Figure 7.7.

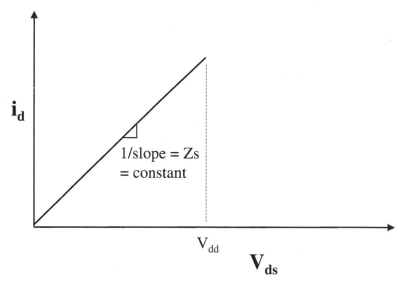

FIGURE 7.7 Ideal *I–V* curve with a fixed impedance.

Recall that the impedance in the saturation region is very high; thus, operation in this region should be avoided when matched impedance is required. If the buffer is designed for operation primarily within the triode region, the impedance will vary much less, as shown in Figure 7.8. Generally, highly linear behavior of the buffer is a desired feature. In early stages of design, when actual buffer curves have not yet been defined, a linear model of the buffer can be used. The information gained from this model can be simulated with a larger system and help in defining the characteristics of the buffer-to-be. In the next section we detail the linear approximation of a buffer.

Let us consider the effect on total impedance of the series resistor (either integrated or discrete) as shown in Figure 7.1. The variation (with process, temperature, etc.) of the *I–V* curve of the transistor will vary. This will, of course, affect the impedance. Figure 7.9 shows three curves for a transistor at a given V_{gs}. This variation should be accounted for. Good estimates may be obtained from silicon designers as to the total variation of a given buffer type. Otherwise, assumptions can be made at this point and system simulations can be conducted to determine the allowable constraints. Constraints on the buffer curves for a given design may ultimately be derived from these simulations and communicated back to buffer designers as design rules.

To see the effect of a series resistor on the buffer, consider Figure 7.10, which shows an N device in series with the resistor. The conclusions derived here will obviously apply to a P device as well. For the total effect of the N device in series with the resistor, the constant impedance of the resistor

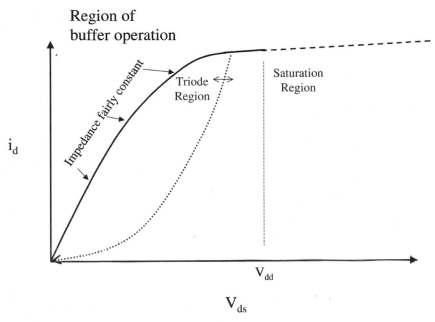

FIGURE 7.8 Different regions of the *I–V* curve exhibit different impedance values.

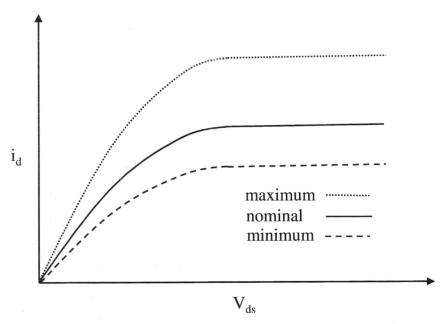

FIGURE 7.9 Variations in the buffer impedance at a constant V_{gs} due to fabrication variations and temperature.

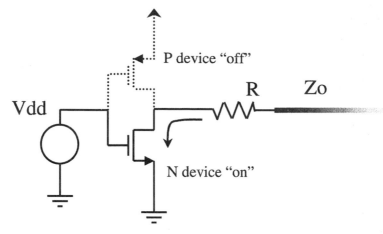

FIGURE 7.10 Buffer in series with a resistor.

must be added to the impedance of the transistor. Since impedance is the inverse of the slope of the transistor curve, the extra impedance of the resistor results in a reduction of the slope of the effective buffer curve, as shown in Figure 7.11. The larger the resistor, the higher the impedance of the buffer and the more linear the curve. For resistors that are large compared to the low-impedance portion of the transistor, the impedance will be dominated by the value of the resistor and the slope of the I–V curve will be approximately $1/R$. Note that with any resistor in series with a buffer, the I–V curve will asymptote at the saturation current of the transistor, since in this condition the transistor is a very high impedance and will dominate the slope of the curve.

7.2.2. Linear Modeling of the CMOS Buffer

During the initial phase of a system design, the characteristics of the driving buffers may not be defined. Since the properties of the buffers are one aspect of a design with many codependent variables, some assumptions must be made about the buffers in order to proceed with the design. Typically, a model with a linear I–V curve is chosen for early simulations rather then "guessing" at a more complex behavioral model such as a transistor curve. The linear I–V model is usually appropriate for this stage of design, for three reasons. First, it is far easier to do multivariable "sweeps" with linear models, due to the simplicity of the model (see Chapter 9). Second, the linear model leaves nothing to wonder about strange effects from a particular choice of buffer model. This can simplify matters greatly. Third, the linear model has been shown to be fairly accurate when used correctly, so more complex models are not motivated early in the design. Requirements for edge rate and driver strength for the system being defined are easily

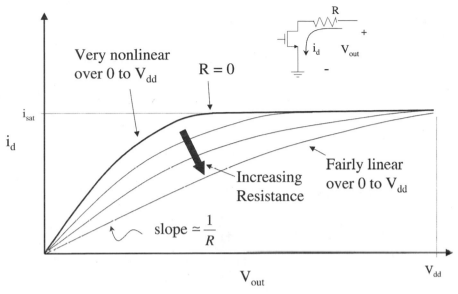

FIGURE 7.11 Effect of increasing the series resistor.

examined during simulation with linear models, and results may easily be tested. If buffers from an already existing driver are to be used, full models from these buffers should be used instead of linear models. Also, if it is known that a buffer will be used that does not have an approximately linear *I–V* curve in the region of interest, a linear buffer model is obviously not merited.

The linear model has two features. It has a *I–V* curve for a pull-up and a pull-down. (Ideally, in a CMOS design, the pull-up and pull-down devices will be very similar.) The model also has a curve defining its switch behavior versus time. The *I–V* curves are a straight-line (resistor) approximation to a transistor curve. How to pick a good linear fit to a transistor curve will be shown later.

The most simplistic linear model for a CMOS buffer is constructed of a switch, a pull-up resistor, and a pull-down resistor. The values of the resistors are chosen to approximate the impedance of the NMOS and PMOS devices in the intended region of operation. Pull-up and pull-down transitions are made with the switch. The major drawback of this model is that there is usually no edge rate control, because the switch is either in one position or the other, which is a step function. In this case, the edge rates of the model are usually governed by the time constant formed by the output capacitance and the parallel combination of the buffer resistance and the characteristic impedance of the transmission line. Figure 7.12*a* shows this simplistic model. This model is generally only used for "back of the envelope" calculations.

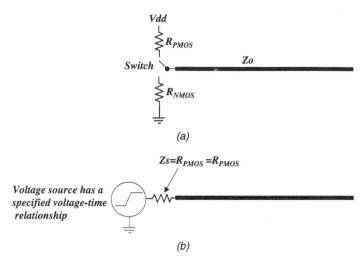

(a)

(b)

FIGURE 7.12   (*a*) Most simplistic linear model of a CMOS buffer; (*b*) another simplistic linear approach.

Another, rather simplistic linear model is shown in Figure 7.12*b*. This model assumes that the impedance of the pull-up and the pull-down are both equal to the impedance Z_s. This modeling approach has the advantage of voltage-time, edge-rate control. The voltage source, V_s, can be specified in SPICE-like simulators to transition from 0 to V_{dd} in a specified amount of time. Furthermore, most simulators allow the creation of a piecewise-linear voltage-time transition so that specific edge shapes can be mimicked.

A more complete linear model would assume separate linear *I–V* curves for the pull-up and pull-down devices and have separate curves that describe how the voltage sources will switch from low to high and from high to low. In this book, these curves are known as switch-time curves. An even more complete model would include two switch-time curves for the pull-up device and two switch-time curves for the pull-down device. This is because each device must be turned on and turned off. An example model for a pull-up and a pull-down device of a basic CMOS buffer is shown in Figure 7.13. The curves shown in Figure 7.13 are input into the simulator. Some simulators assume an edge shape and require the user to input only the impedance and the turn-on and turn-off times of the pull-up and the pull-down. Although various simulators may interpret these curves differently, they are often interpreted as explained below.

Generally, the switch-time curve defines how quickly the impedance, R_{out}, approaches the value in the *I–V* curve when the device is turning on, as shown in Figure 7.14. Basically, the switch-time curve defines how quickly the switch turns on to its full value. When the model is interpreted in this fashion, it will be referred to here as a linear–behavioral model. As shown in Figure 7.15, the varying impedance of the device causes the instantaneous operating point

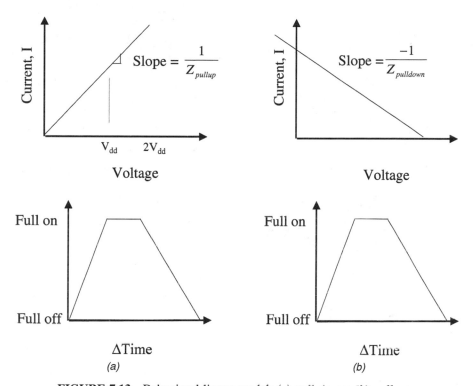

FIGURE 7.13 Behavioral linear model: (*a*) pull-down; (*b*) pull-up.

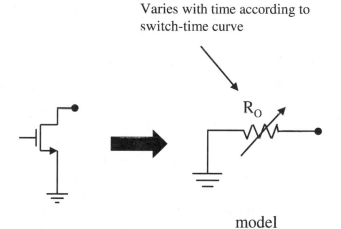

FIGURE 7.14 Basic functionality of linear–behavioral model.

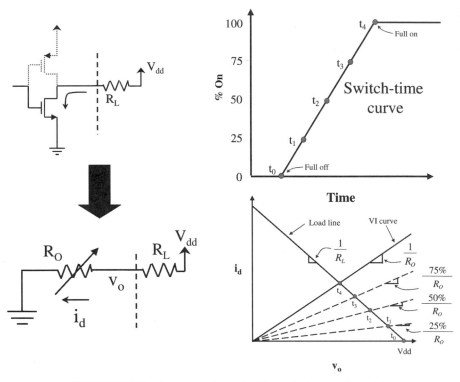

FIGURE 7.15 Interpretation of a linear–behavioral model.

to "climb" up the load curve. This is similar to the behavior of a transistor as V_{gs} increases and the device is transitioning to successively larger, lower impedance V_{gs} curves as the gate voltage increases.

It should be mentioned that some simulators might interpret the I–V and switch-time curves as the model shown in Figure 7.16. In the figure, the switch-time curve is interpreted as a voltage-time curve and is included in the model as a voltage source. The impedance of the model is fixed as the impedance of the I–V curve input into the model. When the model is interpreted in this fashion, it will be referred to here as the linear–linear model. The linear–linear model is likely to be the model used internal to the simulator if the simulation is run in the frequency domain rather than in time domain. To compare this with the variable impedance linear behavioral model, compare Figures 7.15 and 7.17. It can be seen that the location of the points labeled t_0 through t_4 will be slightly different depending on whether the model is interpreted as a linear–linear model or a linear–behavioral model. Also, the linear–linear model can give different results than the linear–behavioral model when considering termination issues. For instance, if $R_o = Z_o$, a reflection terminating at the driver will be matched according to the linear–linear model even when the driver is transitioning. For the linear–behavioral model, the im-

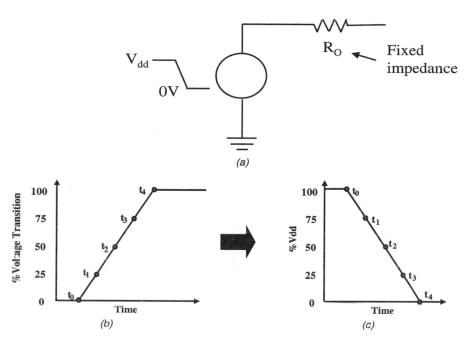

FIGURE 7.16 Basic functionality of linear–linear model: (*a*) circuit; (*b*) switch-time curve; (*c*) pull-down curve.

pedance during the switching interval is dynamic, and thus a reflection arriving during a transition may not be perfectly terminated.

The straight line used to approximate a transistor curve for linear models should be constructed as shown in Figure 7.18. The line is defined by two points, one point the zero current point and the other the point where the transistor curve is expected to cross $V_{dd}/2$, which is generally the threshold voltage at which receivers define the time location of the edge. Also, as shown, lines should be constructed for the nominal, minimum, and maximum expected curve characteristics to account for variation with temperature, silicon process, and so on. Often, prior to the design of the buffer, the system engineer will perform many simulations with linear buffers that will place limitations on the amount of variation, as we elaborate in Chapter 9.

Limitations of the Linear Buffer Model. Although the linear buffer is generally intended for use only in the initial stages of a design cycle, certain limitations should be kept in mind. First, as mentioned before, if a buffer with very nonlinear behavior will be implemented in the final design, the linear buffer will yield very inaccurate results, due to the inability to fit a line to the behavior of the buffer. Second, actual edge shapes and rise and fall times simulated with the linear buffer will not yield the exact results that will be expected in the final system. Finally, errors induced by the use of the

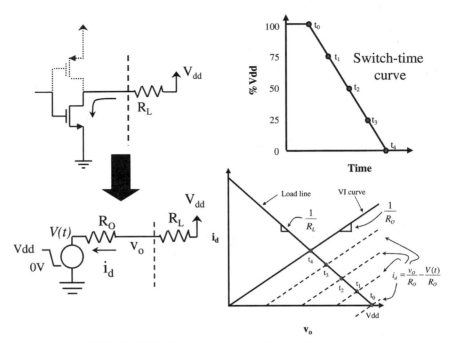

FIGURE 7.17 Interpretation of a linear–linear model.

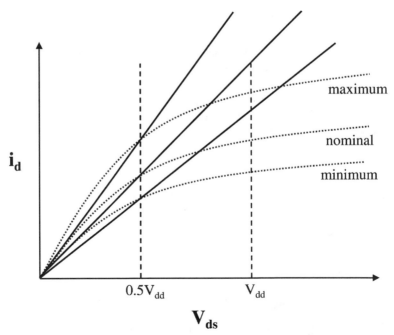

FIGURE 7.18 Measuring the impedance of an *I–V* curve.

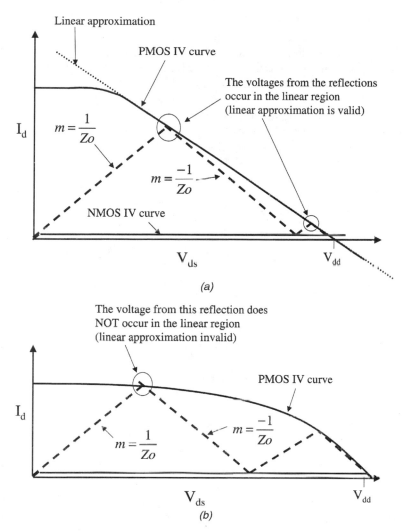

FIGURE 7.19 Using a Bergeron diagram to judge the accuracy of a linear buffer approximation: (*a*) valid case; (*b*) invalid case.

linear model can go unnoticed until very late in the design cycle, when more complex models are used.

To determine the accuracy of the linear approximation, a Bergeron diagram may be used (see Chapter 2). Figure 7.19*a* and *b* show a Bergeron diagram of a CMOS output driver in a high state (the PMOS is on). If the buffer is driving a transmission line with an impedance of Z_o, the Bergeron diagram will predict the voltages at the source due to the reflections on the net. The validity of a linear model approximation can be judged by observing the intersections on the $I–V$ curve. Figure 7.19a shows an example where the voltages from the reflections always intersect the $I–V$ curve in the linear portion. Subsequently, the linear $I–V$ curve (shown as a dotted line) will be adequate when the buffer is driving a transmission line with an impedance of Z_o. Figure 7.19b shows an example where a linear approximation is not adequate because some voltages from the reflections will occur on the nonlinear portion of the $I–V$ curve. Obviously, approximating the interconnect with only one impedance (Z_o) may not be realistic because packages and sockets also introduce discontinuities; however, it is adequate for first-order approximations. This technique may also be useful when specifying buffer characteristics to a silicon designer.

7.2.3. Behavioral Modeling of the Basic CMOS Buffer

The models shown here are similar to the linear–behavioral models in Section 7.2.2. The difference is that these models have $I–V$ curves that mimic the transistor curves instead of approximating it with a straight line. Since these models are defined by a transistor curve and a curve for the time behavior, and are not defined by detailed transistor data such as used in a traditional SPICE model, these models are particularly useful when communicating between different companies. This is because a model can be created that does not contain detailed information about silicon processes and other proprietary information.

A behavioral model consists of an $I–V$ curve and a time curve just as in the linear–behavioral models already discussed. This transistor curve should generally be based on known device characteristics. These characteristics may be supplied by the designer or provided by the vendor. Often, early simulations will be performed using one of the linear models in Section 7.2.2 to determine necessary requirements. The results of the linear simulations may be given to circuit designers or vendors as the initial design targets to create real buffers. Then simulated data based on actual transistor designs (or measured data on the devices) can be used to create a behavioral model.

The structure of a behavioral model is shown in Figure 7.20. For each transistor in the model, an $I–V$ curve and a $V–T$ curve is given for both the turn-on and turn-off cycles. The $V–T$ curves are the shape of the actual voltage waveforms driven into an assumed load. The assumed load used in constructing the model does not have to correspond to the physical load that is expected in a system. The simulator takes as input the load used to construct the model and adjusts the model to different load conditions. However, there may be increased accuracy if the load used to construct the model is similar to the load present in the system. Typically, a load of 50 Ω is chosen; however, other loads can be used to construct the model. Since a pull-up device, for instance, has no capacity to pull down the output, the turn-off $V–T$ curve is constructed from assuming a pull-down load even if no such load exists in the system. Figure 7.21 shows some assumed loads used to construct both the $V–T$ curves for both the pull-up and pull-down devices in a CMOS buffer.

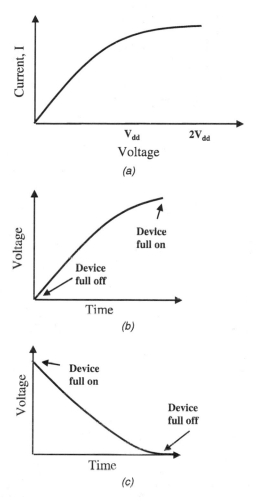

FIGURE 7.20 Requirements of each device (PMOS and NMOS) in a behavioral model: (*a*) *V–I* curve; (*b*) *V–T* curve (turn-on); (*c*) *V–T* curve (turn-off).

Generally, the simulator will construct intermediate *I–V* curves other then the input *I–V* curve to be used during voltage transitions. The intermediate curves will be chosen such that the time behavior during the transition will mimic the *V–T* curve input to the model. This behavior is illustrated in Figure 7.22. Note in the *V–T* curve shown in Figure 7.22 that there is no data point near the threshold (generally, $V_{dd}/2$) at which timing is measured. This may induce a timing inaccuracy, depending on how the simulator (or the model maker) extrapolated points on the *V–T* curve near the threshold. Generally, models should be defined with at least one point at or near the timing threshold.

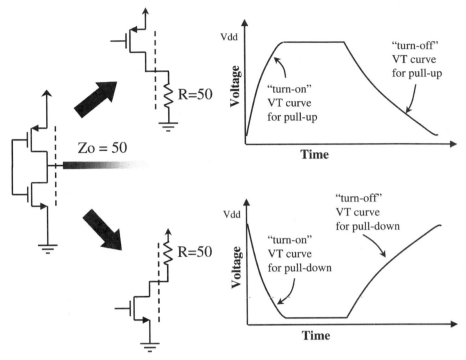

FIGURE 7.21   Creating the $V-T$ curves in a behavioral model.

The $I-V$ and $V-T$ curves used to construct the model can be created from actual silicon measurements or from simulations with full transistor models. In either case, the question arises as to what effects to include in the model. For instance, if a $V-T$ curve is determined directly from measurement of a device that has capacitance present at the output node, the capacitance will obviously affect the shape of the $V-T$ curve measured. This will result in the capacitance being virtually included in the model through its effect on the $V-T$ behavior. If the capacitance under question will actually be present in a physical system, such as on-die capacitance and package capacitance, this has no effect on the driving characteristics of the buffer. However, if the capacitance is embedded in the $V-T$ curve in such a manner, the capacitance will not be modeled properly when the driver node is *receiving* a voltage waveform such as a reflection. For this reason it is preferable to create the $V-T$ curve with as little capacitance as possible at the output node and then add the capacitance back into the final model as a lumped element. Some capacitance cannot easily be subtracted from the model, particularly if the models are created from direct measurement.

As a final note, it should be mentioned that the behavioral model should be constructed with an $I-V$ curve extending in voltage far beyond the expected operation of the device. This accounts for system effects such as a large

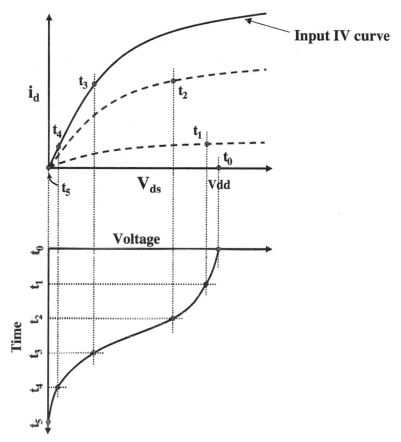

FIGURE 7.22 Example of how the I–V and V–T curves interact.

overshoot. Typically, if the model is anticipated to be used from 0 to V_{dd}, the I–V curve of the model should be extended from $-V_{dd}$ to $2V_{dd}$ to account for worst-case theoretical maximums.

7.3. OUTPUT BUFFERS THAT OPERATE IN THE SATURATION REGION

It was mentioned in Section 7.2 that driver capacitance must be considered carefully when creating a behavioral model, particularly for conditions in which the buffer will receive reflections from the system at the output. Buffers that operate solely in the saturation region of the transistor and thus have very high impedance output are finding growing application in high-speed digital applications. Depending on implementation, these buffers are particularly sensitive to capacitance at the output. If capacitance is modeled incorrectly,

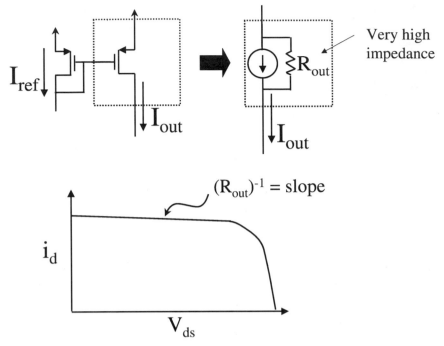

FIGURE 7.23 Example of a buffer that operates in the saturation region.

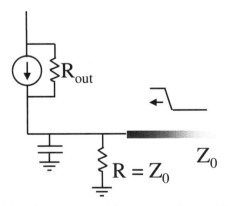

FIGURE 7.24 The output capacitance must be modeled correctly.

simulations can yield drastically incorrect results. The impedance of the output is shown in Figure 7.23.

To see the dependence on output capacitance, consider Figure 7.24. If a high-impedance buffer is terminated at the source end of the transmission line, the termination will probably be a shunt resistor. Thus the capacitance at the output of the buffer can be connected directly to the transmission line.

Thus, the capacitance at the output node affects two things. First, it affects how fast the voltage on the line changes if the current is shut off completely. In other words, the minimum fall time possible at the output depends on the size of the capacitor. Additionally, the capacitance can have a strong effect on the termination because the capacitance will appear initially as a low-impedance device to a waveform arriving at the driver (until it is charged up), which could make the termination ineffective. There are several methods to contend with both of the items noted above. The point made here is simply that the capacitance must be modeled correctly to avoid misleading results when simulating with this type of buffer. As much capacitance as possible should be removed from the device when creating the $V-T$ curve for a buffer of this type. This capacitance must then be placed back into the model as a lumped element.

7.4. CONCLUSIONS

A wide variety of buffer types are encountered in digital systems. No attempt has been made to catalog the various types here. However, the same basic concepts outlined in the example buffers of this chapter can be used to form linear or behavioral models of any general buffer type. One continuing standard started in the 1990s for behavioral models is known as IBIS (I/O Buffer Information Specification). Information on this model format is widely available. Many simulation tool vendors support this format. Behavioral models can also be created for inputs; however, modeling inputs to buffers is generally not nearly as critical as the drive characteristics.

CHAPTER 8

Digital Timing Analysis

We have now covered everything that is needed to model a signal propagating from one component to another. We have covered the details of predicting signal integrity variations and estimating timing impacts caused by a plethora of nonideal high-speed phenomenon. However, this is not sufficient to properly design a digital system. The next step is to coordinate the system so that the individual components can talk to each other. This involves timing the clocks or component strobes so that they can latch in the data at the correct time so that the setup- and hold-time requirements of the receiving components are not violated.

In this chapter we describe the basic timing equations for common-clock and source synchronous bus architectures. The timing equations will allow the engineer to track each timing component that affects system performance, set design targets, calculate maximum bus speeds, and compute timing margins.

8.1. COMMON-CLOCK TIMING

In a common-clock timing scheme, a single clock is shared by driving and receiving agents on a bus. Figure 8.1 depicts a common-clock front-side bus similar to some personal computer designs (a front-side bus is the interface between the processor and the chipset). This example depicts the case when the processor is sending a bit of data to the chipset. The internal latches, which are located at each I/O cell, are shown. A complete data transfer requires two clock pulses, one to latch the data into the driving flip-flop and one to latch the data into the receiving flip-flop. A data transfer occurs in the following sequence:

1. The processor core provides the necessary data at the input of the processor flip-flop (D_p).
2. System clock edge 1 (clk in) is transmitted through the clock buffer and propagates down a transmission line and latches that data from D_p to the output Q_p at the processor.

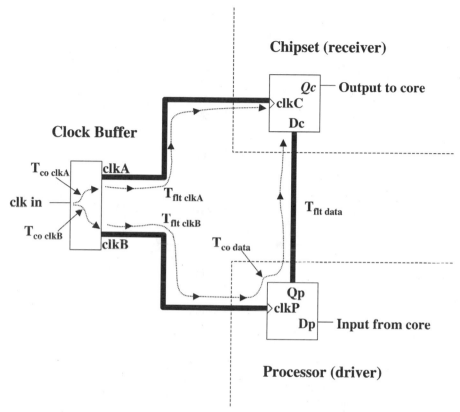

FIGURE 8.1 Block diagram of a common-clock bus.

3. The signal on Q_p propagates down the line to D_c and is latched in by clock edge 2. The data is then available to the core of the chipset.

Based on the foregoing sequence, a few fundamental conclusions can be made. First, the delay of the circuitry and the transmission lines must be smaller than the cycle time. This is because each time a signal is transmitted from one component to another, it requires two clock edges: the first to latch the data at the processor to the output buffer (Q_p), and the second to latch the data at the input of the chipset receiver flip-flop into the core. This places an absolute theoretical limit on the maximum frequency that a common-clock bus can operate. The limitation stems from the total delay of the circuitry and the PCB traces, which must remain less than the delay of one clock cycle. To design a common-clock bus, each of these delays must be accounted for and the setup and hold requirements of the receiver, which are the minimum times that data must be held before and after a clock to ensure correct latching, must be satisfied.

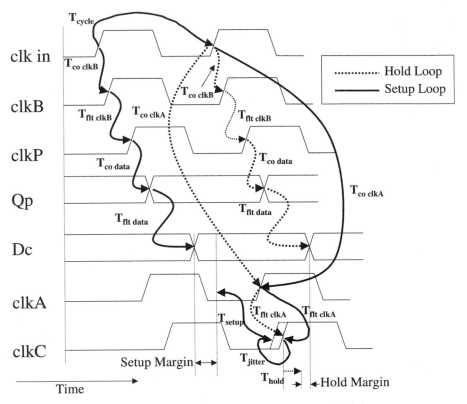

FIGURE 8.2   Timing diagram of a common-clock bus.

8.1.1. Common-Clock Timing Equations

To derive the timing equations for a common-clock bus, refer to the timing diagram in Figure 8.2. Each of the arrows represents a delay in the system and is labeled in Figure 8.1. The solid lines represent the *timing loop* used to derive the equation for the setup margin, and the dashed lines represent the loop used for the hold margin. How to use the timing loop to construct timing equations will become evident shortly.

The delays are separated into three groups: T_{co}'s, flight times, and clock jitter. The T_{co} (time from clock to output) is simply the time it takes for a data bit to appear at the output of a latch or a buffer once it has been clocked in. The *flight times*, T_{flt}, are simply the delays of the transmission lines on the PCB. *Clock jitter*, T_{jitter}, generally refers to the cycle-to-cycle variations of the clock period. Period jitter, for instance, will cause the period of the clock to vary from cycle to cycle, which will affect the timing of the clock edge. For the purposes here, jitter will be considered as a variation that may cause the clock to exhibit a temporary change in clock period.

Setup Timings. To latch a signal into a component, it is necessary that the data signal arrive prior to the clock. The receiver setup time dictates how long the data must be valid before it can be clocked in. In a common-clock scenario, the data are latched to the output of the driver with one clock edge and latched into the receiver with the next clock edge. This means that the sum of the circuit and transmission line delays in the data path must be small enough so that the data signal will arrive at the receiver (D_c) sufficiently prior to the clock signal (clkC). To ensure this, we must determine the delays of the clock and the data signals arriving at the receiver and ensure that the receiver's setup time has been satisfied. Any extra time in excess of the required setup time is the setup margin.

Refer to the solid arrows in the timing diagram of Figure 8.2. The timing diagram depicts the relationship between the data signal and the clocks both at the driver and receiver. The arrows represent the various circuit and transmission line delays in the data and clock paths. The solid arrows form a loop, which is known as the *setup timing loop*. The left-hand portion of the loop represents the total delay from the first clock edge to data arriving at the input of the receiver (D_c). The right-hand side of the loop represents the total delay of the receiver clock.

To derive the setup equation, each side of the setup loop must be examined. First, let's examine the total delay from the first clock edge to data arriving at the input of the receiver. The delay is shown as (see Figure 8.1)

$$T_{\text{data tot}} = T_{\text{co clkB}} + T_{\text{flt clkB}} + T_{\text{co data}} + T_{\text{flt data}} \qquad (8.1)$$

where $T_{\text{co clkB}}$ is the clock-to-output delay of the clock buffer, $T_{\text{flt clkB}}$ the propagation delay of the signal traveling on the PCB trace from the clock chip to the driving component, $T_{\text{co data}}$ the clock-to-output circuit delay of the driver, and $T_{\text{flt data}}$ the propagation delay of the PCB trace from the driver to the receiver.

Now let's examine the total delay of the clock path to the receiver referenced to the first clock edge. This delay is represented by the solid lines on the right-hand side of the setup loop in Figure 8.2. The delay is shown as

$$T_{\text{clock tot}} = T_{\text{cycle}} + T_{\text{co clkA}} + T_{\text{flt clkA}} - T_{\text{jitter}} \qquad (8.2)$$

where T_{cycle} is the cycle time or period of the clock, $T_{\text{co clkA}}$ the clock-to-output delay of the clock buffer, $T_{\text{flt clkA}}$ the propagation delay of the signal traveling on the PCB trace from the clock chip to the receiving component, and T_{jitter} the cycle-to-cycle period variation. The jitter term is chosen to be negative because it produces the worst-case setup margin, as will be evident in the final equation.

The timing margin is calculated by subtracting equation (8.1) from (8.2) and comparing the difference to the setup time required for the receiver. The difference is the setup time margin:

$$T_{\text{setup margin}} = (T_{\text{clock tot}} - T_{\text{data tot}}) - T_{\text{setup}} \qquad (8.3)$$

To design a system, it is useful to break equation (8.3) into circuit and PCB delays, as in equations (8.4) through (8.8)

$$T_{\text{setup margin}} = T_{\text{cycle}} + T_{\text{co clkA}} + T_{\text{flt clkA}} - T_{\text{jitter}} - T_{\text{co clkB}}$$

$$- T_{\text{flt clkB}} - T_{\text{co data}} - T_{\text{flt data}} - T_{\text{setup}} \tag{8.4}$$

The output clock buffer skew is defined as

$$T_{\text{clock skew}} = T_{\text{co clkB}} - T_{\text{co clkA}} \tag{8.5}$$

This is usually specified in the component data sheet. The PCB flight-time skew for the clock traces is defined as

$$T_{\text{PCB skew}} = T_{\text{flt clkB}} - T_{\text{flt clkA}} \tag{8.6}$$

Subsequently, the most useful form of the setup margin equation is

$$T_{\text{setup margin}} = T_{\text{cycle}} - T_{\text{PCB skew}} - T_{\text{clock skew}} - T_{\text{jitter}} - T_{\text{co data}} - T_{\text{flt data}} - T_{\text{setup}}$$

$$\tag{8.7}$$

A common-clock design will function correctly only if the setup margin is greater than or equal to zero. The easiest way to compensate for a setup timing violation is to lengthen the clock trace for the receiver, shorten the clock trace to the driver, and/or shorten the data trace between the driver and receiver flip-flop.

Hold Timings. To latch a signal into the receiver successfully, it is necessary that the data signal remain valid at the input pin long enough to ensure that the data can be clocked in without error. This minimum time is called the *hold time*. In a common-clock bus design, all the circuit and transmission line delays must be accounted for to ensure proper timing relationships to satisfy the hold-time requirements at the receiver. However, even though the second edge in a common clock transaction latches in the data at the receiver flip-flop, it also initiates the next transfer of information by latching data to the output of the driver flip-flop. Subsequently, the hold timing equations must also ensure that the valid data are latched into the receiver before the next data bit arrives. This requires that the delay of the clock path plus the component hold time be less than the delay of the data path. It is basically a race to see which signal can arrive at the receiver first, the new data signal or the clock signal.

To derive the hold-time equation, refer to the dashed arrows in Figure 8.2. The delay of the receiver clock and the delay of the next data transaction must be compared to ensure that the data are properly latched into the receiver before the next data signal arrives at the receiver pin. The clock and data delays are calculated as

$$T_{\text{data delay}} = T_{\text{co clkB}} + T_{\text{flt clkB}} + T_{\text{co data}} + T_{\text{flt data}} \tag{8.8}$$

$$T_{\text{clock delay}} = T_{\text{co clkA}} + T_{\text{flt clkA}} \tag{8.9}$$

Notice that neither the cycle time nor the clock jitter are included in equation (8.9). This is because the hold time does not depend on the cycle time, and clock jitter is defined as cycle-to-cycle period variations. Since the clock cycle time is not needed to calculate hold margin, jitter is not included.

The subsequent hold-time margin is calculated as

$$T_{\text{hold margin}} = (T_{\text{data delay}} - T_{\text{clock delay}}) - T_{\text{hold}} \tag{8.10}$$

If the substitutions of equations (8.5) and (8.6) are made, the useful equation for design is

$$T_{\text{hold margin}} = T_{\text{co data}} + T_{\text{flt data}} + T_{\text{clock skew}} + T_{\text{PCB skew}} - T_{\text{hold}} \tag{8.11}$$

RULES OF THUMB: Common-Clock Bus Design

- Common-clock techniques are generally adequate for medium-speed buses with frequencies below 200 to 300 MHz. Above this frequency, other signaling techniques, such as source synchronous (introduced in the next section), should be used.

- The component delays and the delays of the PCB traces place a hard theoretical limit on the maximum speed a common-clock bus can operate. Subsequently, a maximum limit is placed on the lengths of the PCB traces.

- Trace propagation delays are governed by trace length. Trace lengths are often governed by the thermal solution. As speeds increase, heat sinks get larger and force components farther away from each other, which limit the speed of a common-clock bus design.

8.2. SOURCE SYNCHRONOUS TIMING

Source synchronous timing is a technique where the strobe or clock is sent from the driver chip instead of a separate clock source. The data bit is transmitted to the receiver, and a short time later, a strobe is sent to latch the data into the receiver. Figure 8.3 depicts an example of a source synchronous bus.

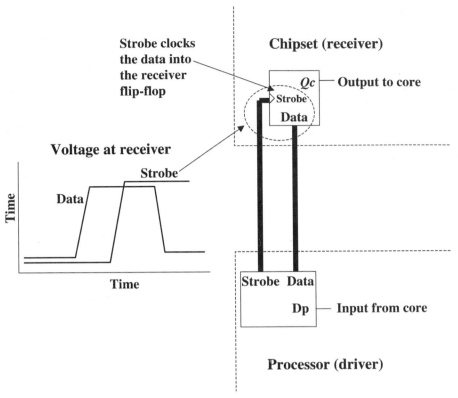

FIGURE 8.3   Relationship between data and strobe in a source synchronous clock bus.

This has several advantages over a common clock. The major benefit is a significant increase in maximum bus speed. Since the strobe and data are sent from the same source, flight time is theoretically no longer a consideration in the equation. Unlike a common clock, where the maximum bus frequency is governed by the circuit and transmission line delays, there is no theoretical frequency limit on a source synchronous bus. There are, however, many practical frequency limitations, which are a manifestation of all the nonideal effects discussed throughout this book. To understand the limitations of a source synchronous bus, it is important to keep in mind that the setup and hold requirements of the receiver must still be met to ensure proper operation. For example, assume that the data are transmitted onto the PCB trace 1 ns prior to the strobe and that the receiver requires a 500 ps setup window. As long as the delay of the data signal is not more than 500 ps longer than that of the strobe, the signal will be captured at the receiver. Thus, the source synchronous design depends on the delay difference between the data and strobe rather than on the absolute delay of the data signal as in common-clock signaling. The difference in the data and strobe delays depend on numerous

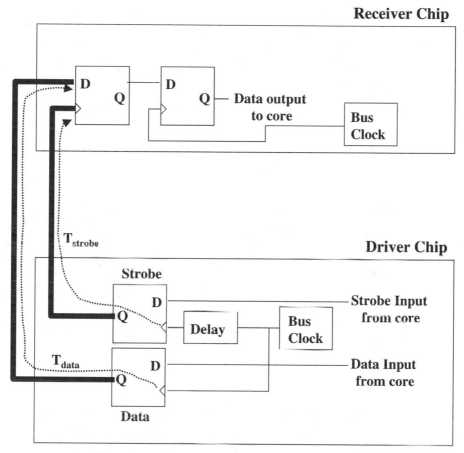

FIGURE 8.4   Block diagram of a source synchronous bus.

factors, such as simultaneous switching noise, trace lengths, trace impedance, signal integrity, and buffer characteristics.

Figure 8.4 shows a block diagram of the circuitry and the source synchronous timing path. The timing path starts at the flip-flop of the transmitting agent and ends at the flip-flop of the receiving agent. Note that the strobe signal is used as the clock input of the receiver flip-flop. The input to the driver flip-flops is generated by the core circuitry. The strobe pattern is usually produced from an internal state machine. The bus clock is generated from a phase-locked loop (PLL) and is usually a multiple of the system clock such as the clock driver typically found on a computer motherboard. For a source synchronous bus to operate properly, the transmission of the strobe must be timed so that both setup and hold requirements of the receiver latch are satisfied. The delay cell in the block diagram achieves this. The delay can be implemented in several ways. Sometimes a state machine is used to clock the

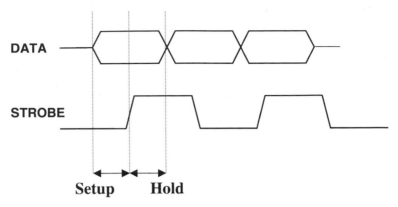

FIGURE 8.5   Setup and hold times in a source synchronous bus.

data flip-flop on one bus clock pulse and the strobe on the next. Other times the data is clocked off the rising edge of the bus clock and the strobe is clocked off the falling edge. The delay cell shown in the block diagram is simply the most generic way of depicting the required offset between data and strobe.

The ideal delay between data and strobe signals depends on the specific circuitry. Typically, however, the ideal offset is 90° assuming a 50% duty cycle. Figure 8.5 depicts a typical data and strobe relationship in a source synchronous bus design.

8.2.1. Source Synchronous Timing Equations

To derive the timing equation for a source synchronous bus, it is necessary to calculate the difference in delay for the data and the strobe path. Figure 8.6 depicts a timing diagram for the simplest implementation of a source synchronous bus. In this particular example, each data transaction requires two clock pulses, the first to clock the data flip-flop at the driver and the second to clock the strobe. Again, both the setup and hold margins must be greater than or equal to zero to guarantee proper timings between components.

Setup Timings. To derive the timing equations, the delay of the data and strobe paths must be calculated as

$$T_{\text{data}} = T_{\text{co data}} + T_{\text{flt data}} \tag{8.12}$$

$$T_{\text{strobe}} = T_{\text{co strobe}} + T_{\text{flt strobe}} + T_{\text{delay}} \tag{8.13}$$

where T_{delay} is the offset between data and strobe. The setup margin is calculated by subtracting (8.12) from (8.13) and comparing the difference to the

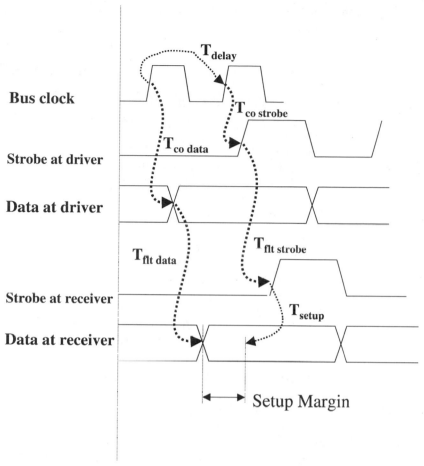

FIGURE 8.6 Setup timing diagram for a source synchronous bus.

receivers required setup time as in

$$T_{\text{setup margin}} = (T_{\text{co strobe}} + T_{\text{flt strobe}} + T_{\text{delay}}) - (T_{\text{co data}} + T_{\text{flt data}}) - T_{\text{setup}}$$

$$(8.14)$$

where $T_{\text{co strobe}}$ is the clock-to-output delay of the strobe flip-flop, $T_{\text{flt strobe}}$ the propagation delay of the strobe PCB trace from the driver to receiver, $T_{\text{co data}}$ the clock-to-output delay of the driver flip-flop, $T_{\text{flt data}}$ the propagation delay of the data PCB trace from the driver to the receiver, and T_{delay} the delay between data and strobe clocking. In this example, the delay is one clock period. It is left as a delay so that the equations will be generic. The timing diagram is shown in Figure 8.6.

To simplify the equation, a few terms need to be defined.

$$T_{vb} = T_{co\ data} - (T_{co\ strobe} + T_{delay})$$ (8.15)

$$T_{PCB\ skew} = T_{flt\ data} - T_{flt\ strobe}$$ (8.16)

T_{vb}, the "valid before" time, refers to the time before the strobe occurs when the data will be valid. $T_{PCB\ skew}$ is the difference in flight times between data and strobe signals. Note that this term actually represents the total skew from silicon pad at the driver to silicon pad at the receiver, including all packages, sockets, and every other attribute that can change the delay of the signal. It is not only the delay skew due to PCB trace mismatches; do not be confused by the name.

The simplified equation for setup margin is

$$T_{setup\ margin} = -T_{vb} - T_{setup} - T_{PCB\ skew}$$ (8.17)

Note that the quantity T_{vb} is negative. This is because the standard way of calculating skews in a source synchronous design is data minus strobe, and to satisfy setup requirements, the data must arrive at the receiver prior to the strobe. Subsequently, in a working design, (8.15) will always be negative. The negative sign in (8.17) is required to calculate a positive margin.

Hold Timings. The hold-time margin is calculated in the same manner as the setup time, except that we compare the delay between the first strobe cycle and the second data transition. Subtracting the delay of the strobe from the delay of the second data transition and comparing the difference to the receivers hold-time requirement derives

$$T_{hold\ margin} = (T_{co\ data} + T_{flt\ data} + T_{delay}) - (T_{co\ strobe} + T_{flt\ strobe}) - T_{hold}$$
(8.18)

The timing diagram is shown in Figure 8.7. Again, to simplify the equation to a useful form, equations (8.19) and (8.16) are used. T_{va} is the "valid after" time and refers to the time after the strobe for which the data signal is still valid.

$$T_{va} = T_{co\ data} - T_{co\ strobe} + T_{delay}$$ (8.19)

The simplified equation for hold margin is

$$T_{hold\ margin} = T_{va} - T_{hold} - T_{PCB\ skew}$$ (8.20)

Again, note that the components of the above equations represent the total skew from silicon pad at the driver to silicon pad at the receiver, including

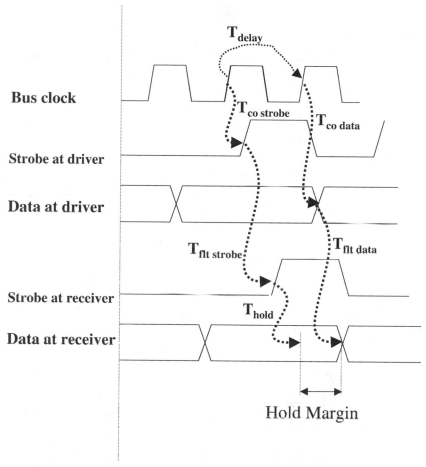

FIGURE 8.7   Hold timing diagram for a source synchronous bus.

all packages, sockets, and all other attributes that can change the delay of the signal.

8.2.2. Deriving Source Synchronous Timing Equations from an Eye Diagram

A convenient graphical method to analyze timings is known as an *eye diagram*. Figure 8.8 depicts an idealized eye diagram of the data and strobe at the receiver. It is easy to equate T_{va} and T_{vb} to the sum of the skew, hold/setup time, and the margins. The components can then be rearranged to yield the proper equations. This point of view gives more insight into source synchronous

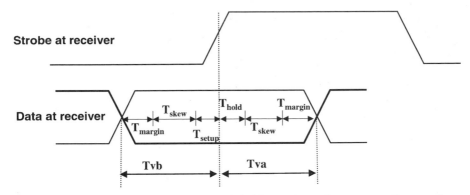

FIGURE 8.8 Calculating the setup and hold margins using an eye diagram for a source synchronous bus.

timings. The equations derived from the eye are identical to those derived from the timing diagrams. Equations (8.22) and (8.24) are the final equations for source synchronous setup and hold margin. Note that the sign convention of T_{vb} in equation (8.24) is reversed from the sign convention used in (8.17). This is done to better fit the graphical representation as applied in the eye diagram.

$$T_{va} = T_{hold\ margin} + T_{hold} + T_{PCB\ skew} \qquad (8.21)$$

$$T_{hold\ margin} = T_{va} - T_{hold} - T_{PCB\ skew} \qquad (8.22)$$

$$T_{vb} = T_{setup\ margin} + T_{setup} + T_{PCB\ skew} \qquad (8.23)$$

$$T_{setup\ margin} = T_{vb} - T_{setup} - T_{PCB\ skew} \qquad (8.24)$$

RULES OF THUMB: Source Synchronous Timings

- There is no theoretical limit on the maximum bus speed.
- Bus speed is a function of the difference in delay (skew) between data and strobe.
- Nonideal effects cause unwanted skew and thus place practical limits on the speed of a source synchronous bus.
- Flight time is not a factor in a source synchronous bus.
- It is beneficial to make the strobe signal identical to the data signal. This will minimize skew.

It should be noted that every single effect so far covered in this book could affect the delays or skews of the signals. Simultaneous switching noise, non-ideal return paths, impedance discontinuities, ISI, connectors, packages, and any other nonideal effect must be included in the analysis.

8.2.3. Alternative Source Synchronous Schemes

There are several alternative source synchronous schemes. Many of them significantly increase the bus clock by multiplying the system clock. Figure 8.9 is an example where the system clock is multiplied by a factor of $2\times$ and a dual-strobe methodology is used to clock in the data. In this particular scheme, the data are generated from the rising edge of the bus clock and the strobe is

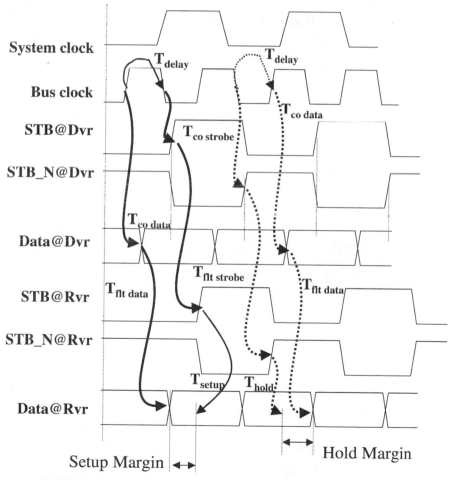

FIGURE 8.9 Alternate timing sequence for a source synchronous bus.

generated from the falling edge. Alternating rising edges of STB (strobe) and
STB_N (inverse strobe) clock in each block of data. That is, data block 1 is
clocked in by the rising edge of STB, and data block 2 is clocked in by the
rising edge of STB_N. The equations derived in Section 8.2.1 apply.

8.3. ALTERNATIVE BUS SIGNALING TECHNIQUES

As speeds increase, source synchronous timing is getting ever more difficult to
implement. It gets significantly more difficult to control skew as frequencies
increase. Nonideal effects such as simultaneous switching noise, nonideal return paths, intersymbol interference, and crosstalk increase skew dramatically.
Furthermore, any sockets or connectors also introduce additional variables that
could increase skew. As mentioned earlier, it is optimal to make the data and
the strobe paths identical to each other so that they look the same at the receiver. If the nets are identical (or at least close to identical), the timing and
signal integrity differences will be minimal, subsequently, the skew will be
minimized. One problem with conventional source synchronous timing techniques is that the strobe is sent out a significant amount of time after the data
(usually several hundred pico-seconds to a few nano-seconds). During this
time, noise from the core, power delivery or other parts of the system can be
coupled onto the strobe, varying its timing and signal quality characteristics
so that they differ from the data. This increases the skew significantly.

New bus signaling techniques that minimize the effect of skew are being
developed continuously. Here are a few alternative techniques.

8.3.1. Incident Clocking

In the technique of incident clocking, the data and strobe are sent out simultaneously instead of separated by a delay as in conventional source synchronous
timings. This allows the data and the strobe to experience the same coupled
noise and will subsequently be subjected to similar timing push-outs and signal
integrity distortions. The result will be a decrease in the skew at the receiver
and subsequently an increase in the maximum speed. But if the data and strobe
are sent out simultaneously, how are the setup- and hold-time requirements of
the receiver satisfied? The only obvious way is to delay the strobe on the silicon at the receiver; however, this partially defeats the original intent of incident
strobing because noise can be coupled into the strobe in the receiving circuitry
during the strobe delay. Theoretically, however, the coupled noise should be
significantly smaller than in conventional source synchronous architecture.

8.3.2. Embedded Clock

Another promising alternative to source synchronous timing is to embed the
clock into the data signal by borrowing techniques from the communication

industry. This technique would eliminate the need for a separate strobe. In this technique, a PLL constructs a clock from the data patterns themselves. However, since the PLL requires some minimum data switching in order to construct a clock, there is some overhead in maintaining sufficient data signals. For example, if the data to be transmitted consist of a long string of 0's, the algorithm must send periodic 1's to keep the PLLs in the driver and receiver in phase. Although this technique sounds promising, it is estimated that the algorithm would require approximately a 20% overhead; that is, for every 8 bits of data transmitted, two clocks are transmitted.

CHAPTER 9

Design Methodologies

Digital design is entering a new realm. Bus speeds have increased to a point where high-frequency phenomena that previously had second- or third-order effects on system performance have become first order. Many of these numerous high-speed issues have never needed to be considered in digital design of times past. Subsequently, modern high-speed design not only requires the engineer to continuously break new ground on a technical level, but it also requires the engineer to account for significantly more variables. Since the complexity of a design increases dramatically with the number of variables, new methodologies must be introduced to produce a robust design. The methodologies introduced in this chapter show how to systematically reduce a problem with an otherwise intractable amount of variables to a solvable problem.

Many previous designs have used the route–simulate–route method. This old methodology requires the layout engineer to route the board prior to any simulation. The routed board is then extracted and simulated. When timing and/or signal integrity errors are encountered, the design engineer determines the necessary corrections, the layout is modified, and the loop begins all over again. The problem with the old methodology is that it takes a significant amount of time to converge and often does not provide a thorough understanding of the solution space. Even when a solution is determined, there may not be an understanding of why the solution worked. A more efficient method of design process would entail a structured procedure to ensure that all variables are accounted for in the pre-layout design. If this is done correctly, a working layout will be produced on the first pass and then the board extraction and simulation process is used only to double check the design.

The methodologies introduced in this section concentrate on efficiently producing high-speed bus designs for high-volume manufacturing purposes. Additionally, it outlines proven strategies that have been developed to handle the large number of variables that must be accounted for in these designs. Some of the highest-performance digital designs on the market today have been developed using these or a variation of the methodologies in this chapter. Finally, in this section we introduce a specific design flow that allows the engineer to proceed from the initial specifications to a working bus design with minimal layout iteration. This methodology will produce robust digital designs, improve time to market, and increase the efficiency of the designer.

9.1. TIMINGS

The only thing that really matters in a digital design is timings. Some engineers argue that cost is the primary factor; however, the cost assumption is built into most designs. If the design does not meet the specific cost model, it will not be profitable. If the timing equations presented in Chapter 8 are not solved with a positive margin, the system simply will not work. Every concept in this book somehow relates to timings. Even though several chapters in this book have concentrated on such things as voltage ringback or signal integrity problems, they still relate to timings because signal integrity problems matter only when they affect timings (or destroy circuitry, which makes the timing equations unsolvable). This chapter will help the reader relate all the high-speed issues discussed in this book back to the equations presented in Chapter 8.

The first step in designing a digital system is to roughly define the initial system timings. To do so, it is necessary to obtain first-order estimates from the silicon designers on the values of the maximum and minimum output skew, T_{co}, and setup and hold times for each component in the system. If the silicon components already exist, this information will usually be contained in the data sheet for the component. A spreadsheet should be used to implement the digital timing equations derived in Chapter 8 or the appropriate equations that are required for the particular design. The timing equations are solved assuming a certain amount of margin (presumably zero or a small amount of positive margin). Whatever is left over is allocated to the interconnect design. If there is insufficient margin left over to design the interconnects, either the silicon numbers need to be retargeted and redesigned, or the system speed should be decreased.

The maximum and minimum interconnect delays should initially be approximated for a common-clock design to estimate preliminary maximum and minimum length limits for the PCB traces and to ensure that they are realistic. If the maximum trace length, for example, is 0.15 in., it is a good bet that the board will not be routable. To do this, simply set the setup and hold margins to zero (using the equations in Chapter 8), solve for the trace delays, and translate the delays to inches using an average propagation speed in FR4 of 150 ps/in. for a microstrip or 170 ps/in. for a stripline. Of course, if the design is implemented on a substrate other than FR4, simply determine the correct propagation delay using equation (2.3) or (2.4) using an average value of the effective dielectric constant. In a source synchronous design, the setup and hold margins should be set to zero and the PCB skew times should be estimated. Again, the skews should be checked at this point to ensure that the design is achievable.

It should be noted that every single value in the spreadsheet is likely to change (unless the silicon components are off-the-shelf parts). The initial timings simply provide a starting point and define the initial design targets for the interconnect design. They also provide an initial analysis to determine whether or not the bus speed chosen is realistic. Typically, as the design progresses, the silicon and interconnect numbers will change based on laboratory testing and/or more detailed simulations.

9.1.1. Worst-Case Timing Spreadsheet

A spreadsheet is not always necessary for a design. If all the components are off the shelf, it may not be necessary to design with a spreadsheet because the worst-case component timings are fixed. However, if the components, such as a processor, chipset, or a memory component, is being developed simultaneously with the system, the spreadsheet is an extremely valuable tool. It allows the component design teams (i.e., the silicon designers) and the system design teams to coordinate with each other to produce a working system. The spreadsheet is updated periodically and is used to track progress, set design targets, and perform timing trade-offs.

The worst-case spreadsheet calculates the timings, assuming that the unluckiest guy alive has built the system (i.e., all the manufacturing variables and the environmental conditions are chosen to deliver the worst possible performance). Statistically, a system with these characteristics may never be built, however, assuming that it will guarantee that the product will be reliable and robust over all process variations and environmental conditions.

Figures 9.1 and 9.2 are examples of worst-case timing spreadsheets for a source synchronous and a common-clock bus design, respectively. Notice that an extra term, T_{guard}, has been added into the equations. This represents the *tester guard band*, which is the accuracy to which the components can be tested in high volume. The tester guard band provides extra margin that is designed into the system to account for inaccuracies in silicon timing measurements which are always present in a high-volume tester. The spreadsheet is constructed so that each timing component can be tracked for each driver and each receiver. The spreadsheets shown only depict a portion of the timings for the case when agent 1 is driving agent 2. The complete spreadsheets

Source Synchronous Agent 1 driving agent 2

Setup

$$-T_{vb,\,\min} \quad -T_{skew,\,\max} \quad -T_{setup} \quad -T_{guard} \quad = T_{setup\ margin}$$

	Tvb	Tpcb skew	Tsetup	Tguard	Tmargin
Target (ns)	1.25	0.55	0.5	0.2	0
Predicted (ns)	1.25	0.65	0.5	0.2	-0.1
Difference (ns)	0	-0.1	0	0	-0.1

Hold

$$T_{va,\min} \quad -T_{skew,\min} \quad -T_{hold} \quad -T_{guard} \quad = T_{hold\ margin}$$

	Tva	Tpcb skew	Thold	Tguard	Tmargin
Target (ns)	1.25	0.55	0.5	0.2	0
Predicted (ns)	1.25	0.53	0.47	0.2	0.05
Difference (ns)	0	-0.05	-0.1	0	0.05

FIGURE 9.1 Example of a source synchronous timing spreadsheet.

Common clock agent 1 driving agent 2

Setup

$$T_{cycle} - T_{PCB\ skew,max} - T_{clock\ skew,\ max} - T_{jitter} - T_{co\ data,\ max} - T_{flt\ data,max} - T_{setup} - T_{guard} = T_{setup\ margin}$$

	Tcycle	Tpcb skew	Tclock skew	Tjitter	Tco data	Tflt data	Tsetup	Tguard	Tmargin
Target (ns)	5	0.5	0.2	0.1	0.5	1.8	0.8	0.75	0.35
Predicted (ns)	5	0.6	0.18	0.19	0.52	1.9	0.78	0.75	0.08
Difference (ns)	0	-0.1	0.02	-0.09	-0.02	-0.1	0.02	0	0.27

Hold

$$T_{co\ data,min} + T_{flt\ data,min} + T_{clock\ skew,min} + T_{PCB\ skew,min} - T_{hold} - T_{guard} = T_{hold\ margin}$$

	Tco data	Tflt data	Tclock skew	Tpcb skew	Thold	Tguard	Tmargin
Target (ns)	0.5	1.8	-0.2	-0.5	0.8	0.75	0.05
Predicted (ns)	0.47	1.6	0.18	0.6	0.82	0.75	-0.28
Difference (ns)	0.03	0.2	0.02	-0.1	-0.02	0	0.33

FIGURE 9.2 Example of a common-clock timing spreadsheet.

would depict the timings for each agent driving. The target row refers to the
initial estimations of each portion of the timing equation prior to simulation.
The target numbers are based primarily on past design experience and "back
of the envelope" calculations. The predicted row refers to the postsimulation
numbers. These numbers are based on numerous simulations performed during the sensitivity analysis, which will be described later in this chapter. The
predicted T_{va}, T_{vb}, jitter, T_{co}, output skew, and setup/hold numbers are based off
the silicon (I/O) designer's simulations and should include all the effects that
are not "board centric." This means that the simulations should not include
any variables under the control of the system designer, such as sockets, PCB,
packaging, or board-level power delivery effects. The PCB skew and the flight
time numbers constitute the majority of this book. These effects include all
system-level effects, such as ISI, power delivery, losses, crosstalk, and any
other applicable high-speed effect. Also note that the timings are taken from
silicon pad at the driver to silicon pad at the receiver. Subsequently, don't get
confused by the terminology *PCB skew*, because that term necessarily includes
any package, connector, or socket effects.

Also note that the signs on the clock and PCB skew terms should be observed for the timing equations. For example, the common-clock equations
indicate that increasing the skew values will help the hold time because they
are positive in the equation. However, the reader should realize that the signs
are only a result of how the skew was defined, as in equations (8.5) and (8.6).
The skews were defined with the convention of data path, minus clock path
and the numbers entered into the spreadsheet should follow the same convention. For example, if silicon simulations or measurements indicate that the
output skew between clkB and clkA (measured as $T_{clkB} - T_{clkA}$) varies from
-100 to $+100$ ps, the minimum value entered into the common-clock hold
equation is -100 ps, and the maximum value entered into the setup equation

is +100 ps. It is important to keep close track of how the skews are measured to determine the proper setup and hold margins. The proper calculation of skews is elaborated further in Section 9.2.

9.1.2. Statistical Spreadsheets

The problem with using a spreadsheet based on the worst-case timings is that system speeds are getting so fast that it might be impossible to design a system that works, assuming that all components are worst case throughout all manufacturing processes. Statistically, the conditions of the worst-case timing spreadsheet may happen only once in a million or once in a billion. Subsequently, in extremely high-speed designs where the worst-case spreadsheet shows negative margin, it is beneficial to use a statistical spreadsheet in conjunction with (*not in place of*) the worst-case timing spreadsheet. The statistical spreadsheet is used to assess the risk of a timing failure in the worst-case spreadsheet. For example, if the worst-case timing spreadsheet shows negative margin but the statistical spreadsheet shows that the probability of a failure occurring is only 0.00001%, the risk will probably be deemed acceptable. If it is shown that the probability of a failure is 2%, the risk will probably not be acceptable.

The statistical spreadsheet requires that a mean and a standard deviation be obtained for each timing component. The spreadsheets are then constructed with the following equations:

$$T_{\text{setup margin}} = \mu_{\text{cycle}} - \mu_{\text{PCB skew}} - \mu_{\text{clock skew}} - \mu_{\text{co data}} - \mu_{\text{flt data}} - \mu_{\text{setup}} - T_{\text{guard}}$$

$$- k\sqrt{\sigma_{\text{cycle}}^2 + \sigma_{\text{PCB skew}}^2 + \sigma_{\text{clock skew}}^2 + \sigma_{\text{co data}}^2 + \sigma_{\text{flt data}}^2 + \sigma_{\text{setup}}^2}$$

$$(9.1)$$

$$T_{\text{hold margin}} = \mu_{\text{co data}} + \mu_{\text{flt data}} + \mu_{\text{clock skew}} + \mu_{\text{PCB skew}} - \mu_{\text{hold}} - T_{\text{guard}}$$

$$- k\sqrt{\sigma_{\text{co data}}^2 + \sigma_{\text{flt data}}^2 + \sigma_{\text{clock skew}}^2 + \sigma_{\text{PCB skew}}^2 + \sigma_{\text{hold}}^2} \qquad (9.2)$$

$$T_{\text{margin, setup}} = -\mu_{\text{vb}} - \mu_{\text{setup}} - \mu_{\text{PCB skew}} - T_{\text{guard}} - k\sqrt{\sigma_{\text{vb}}^2 + \sigma_{\text{setup}}^2 + \sigma_{\text{PCB skew}}^2}$$

$$(9.3)$$

$$T_{\text{margin, hold}} = \mu_{\text{va}} - \mu_{\text{hold}} - \mu_{\text{PCB skew}} - T_{\text{guard}} - k\sqrt{\sigma_{\text{va}}^2 + \sigma_{\text{hold}}^2 + \sigma_{\text{PCB skew}}^2} \qquad (9.4)$$

where μ is the timing mean, σ the standard deviation, and k the number of standard deviations (i.e., $k = 3$ refers to a 3σ spread). Equations (9.1) and (9.2) are for common-clock buses and (9.3) and (9.4) are for source synchronous buses.

It should be noted that these equations assume that the distributions are approximately normal (Gaussian) and that the timing components are completely independent of each other. Technically, neither of these assumptions is correct in a digital design; however, the approximations are close enough to assess the risk of a failure. If the engineer understands these limitations and uses the worst-case spreadsheet as the primary tool, the statistical equations can be a powerful asset. Furthermore, the equations below represent only one way to implement a statistical spreadsheet, a resourceful engineer may come up with a better method. Equations (9.1) through (9.4) can be compared directly to the equations derived in Chapter 8. However, note that there is no jitter term because the jitter is included in the statistical variation of the cycle time. Notice the tester guard band (T_{guard}) term is not handled statistically. This is because the statistical accuracy of the testers is usually not known.

The most significant obstacle in using a statistical timing spreadsheet is determining the correct statistical means and distributions of the individual components. This becomes especially problematic if portions of the system will be obtained from numerous sources. For example, PCB vendors each have a different process, which will provide different mean velocities and impedance values with different distribution shapes. Furthermore, the statistical data for a given vendor may change. Sometimes, however, the statistical behavior can be estimated. For example, consider a PCB vendor which claims that the impedance of the PCB board will have a nominal value of 50 Ω and will not vary more than $\pm 15\%$. A reasonable assumption would be to assume a normal distribution and that the $\pm 15\%$ points are at three standard deviations (3σ). Approximations such as this can be used in conjunction with Monte Carlo simulations to produce statistical approximations of flight time and skew. Alternatively, when uniform distributions are combined in an interconnect simulation using large Monte Carlo analysis, the result usually approximately resembles a normal distribution. Subsequently, the interconnect designer could approximate the statistical parameters of flight time or flight-time skew by running a large Monte Carlo analysis where all the input variables are varied randomly over uniform distributions bounded by the absolute variations. Since the resultant of the analysis will resemble a normal distribution, the mean and standard deviation can be approximated by examining the resultant data. Numerous tools, such as Microsoft Excel, Mathematica, and Mathcad, are well suited to the statistical evaluation of large data sets. It is possible that the statistical technique described here may have some flaws; however, it is adequate to approximate the risk of a failure if the worst-case spreadsheet shows negative margin.

Ideally, the designer would be able to obtain accurate numbers for the mean and standard deviation for all or most of the variables in the simulation (i.e., PCB Z_o, buffer impedance, input receiver capacitance, etc.). Monte Carlo analysis (an analysis method available in most simulators in which many variables are varied fairly randomly and the simulated results are observed) is then performed with a large number of iterations. The resultant timing numbers are then inserted into the statistical spreadsheet. Note that similar analysis is required from the silicon design team to determine the mean and standard deviation of the silicon timings (i.e., T_{vb}, T_{va}, T_{setup}, T_{hold}, and T_{jitter}).

To determine the risk of failure for a design that produces negative margin in a worst-case timing spreadsheet, the value of k is increased from 1 incrementally until the timing margin is zero. This will determine how many standard deviations from the mean timings are necessary to break the design and will provide insight into the probability of a failure. For example, if the value of k that produces zero margin is 1, the probability that the design will exhibit positive margin is only 68.26%. In other words, there is a 31.74% chance of a failure (i.e., a design that produces negative timing margin). However, if the value of k that produces zero margin is 3.09, the probability that the timing margin will be greater or equal to zero is 99.8%, which will produce 0.2% failures. In reality, however, a 3.09σ design will probably produce a 99.9% yield instead of 99.8% simply because the instances that will cause a failure will usually be skewed to one side of the bell curve. When considering the entire distribution, this analysis technique assumes that the probability of a result lying outside the area defined by sigma is equivalent to the percentage of failures. Obviously, this is not necessarily true; however, it is adequate for risk analysis, especially since it leans toward conservatism. Typically, in high-volume manufacturing, a 3.0σ design (actually, 3.09), is considered acceptable, which will usually produce a yield of 99.9%, assuming one-half of the curve.

Figure 9.3 demonstrates how the value of k relates to the probability of an event occurring for a normal distribution. Table 9.1 relates the value of k to the percentage of area not covered, which is assumed to be the percentage of failures.

9.2. TIMING METRICS, SIGNAL QUALITY METRICS, AND TEST LOADS

After creating the initial spreadsheet with the target numbers, specific timing and signal quality metrics must be defined. These metrics allow the calculation of timing numbers for inclusion into the spreadsheet and provide a means of determining whether signal quality is adequate.

9.2.1. Voltage Reference Uncertainty

The receiver buffers in a digital system are designed to switch at the threshold voltage. However, due to several variables, such as process variations and system noise, the threshold voltage may change relative to the signal. This variation in the threshold voltage is known as the V_{ref} uncertainty. This uncertainty is a measure of the relative changes between the threshold voltage and the signal. Figure 9.4 depicts the variation in the threshold voltage (known as the *threshold region*).

Signal quality, flight time, and timing skews are measured in relation to the threshold voltage and are subsequently highly dependent on the size of

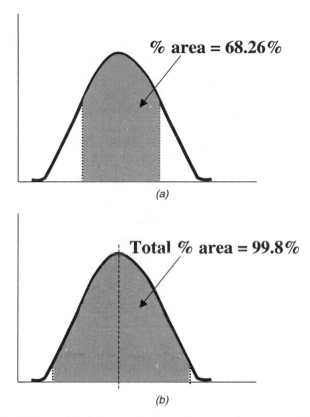

FIGURE 9.3 Relationship between sigma and area under a normal distribution: (*a*) $k = 1$ standard deviation (1σ design); (*b*) $k = 3.09$ standard deviations (3.09σ design).

TABLE 9.1. Relationship Between k and the Probability That a Result Will Lie Outside the Solution

k	1	2	3	3.09	3.72	4.26
Percent failure	31.74	4.56	0.27	0.2	0.02	0.002

Source: Data from Zwillinger [1996].

the threshold region. The threshold region includes the effects on both the reference voltage and the signal. The major effects that typically contribute to the V_{ref} uncertainty are:

- Power supply effects (i.e., switching noise, ground bounce, and rail collapse)
- Core noise from the silicon circuitry
- Receiver transistor mismatches

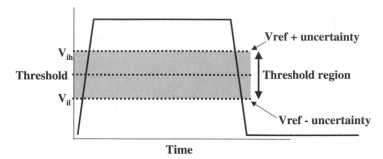

FIGURE 9.4   Variation in the threshold voltage.

- Return path discontinuities
- Coupling to the reference voltage circuitry

The threshold region typically extends 100 to 200 mV above and below the threshold voltage. The upper and lower levels will be referred to as V_{ih} and V_{il}, respectively. The V_{ref} uncertainty is important to quantify accurately early in the design cycle because it accounts for numerous effects that are very difficult or impossible to model explicitly in the simulation environment. Often, they are determined through the use of test boards and test chips.

9.2.2. Simulation Reference Loads

In Section 9.1.1 the concept of a timing spreadsheet is introduced. It is important that all the numbers entered into the spreadsheet (or timing equations) are calculated in such a manner that they add up correctly and truly represent the timings of the entire system. This is done with the use of a standard simulation test load (referred to as either the standard or the reference load). The reference load is used by the silicon designers to calculate T_{co} times, setup/hold times, and output skews. The system designers use it to calculated flight times and interconnect skews. The reference load is simply an artificial interface between the silicon designer world and the system designer world that allows the timings to add up correctly. This is important because the silicon is usually designed in isolation of the system.

If the silicon timing numbers are generated by simulating into a load that is significantly different from the load the output buffer will see when driving the system, the sum of the silicon-level timings and the board-level timings will not represent the actual system timings. For example, Figure 9.5 shows how simulating the T_{co} timings into a load that is different from the system can artificially exaggerate or understate the timings of the system. The top portion of the figure depicts the T_{co} time of an output buffer driven into a load that does not resemble the system input. The bottom portion represents the total delay of a signal in the system (shown as the output buffer driving a transmission

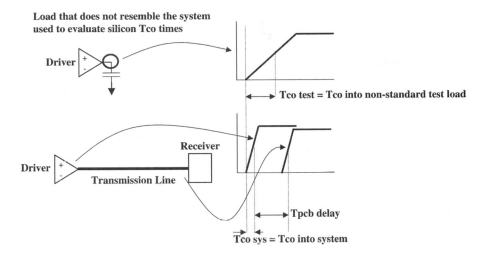

$$Total\ delay = Tco\ sys + Tpcb\ delay \neq Tco\ test + Tpcb\ delay$$

FIGURE 9.5 Illustration of timing problems that result if a standard reference load is not used to insert timings into the spreadsheet.

line). The actual delay the signal experiences between the driver and receiver in the system is $T_{co\ sys} + T_{PCB\ delay}$. If the silicon T_{co} numbers are calculated with a load that does not look electrically similar to the system, then the total delay will be $T_{co\ test} + T_{PCB\ delay}$, which will not equal $T_{co\ sys} + T_{PCB\ delay}$.

To prevent this artificial inflation or deflation of calculated system margins, the component timings should be simulated with a standard reference load. The same load is used in the calculation of system-level flight times and skews. This creates a common reference point to which all the timings are calculated. This prevents the spreadsheets from incorrectly adding board- and silicon-level timings. If this methodology is not performed correctly, the predicted timing margins will not reflect reality. Since the timing spreadsheets are the base of the entire design, such a mistake could lead to a nonfunctional design. In the sections on flight time and flight-time skew we explain in detail how to use this load.

Choosing a Reference Load. Ideally, the reference load should be very similar to the input of the system. It is not necessary, however, that the reference load be electrically identical to the input to the system, just similar enough that the buffer characteristics do not change. Since the standard load acts as a transition between silicon and board timings, small errors will be canceled out. This is demonstrated with Figure 9.6. In Figure 9.6 the reference load is different from the input to the system. However, since the load is used as a reference point, the total delay is calculated correctly when the timings are summed in the spreadsheet. The top waveform depicts the T_{co} as measured

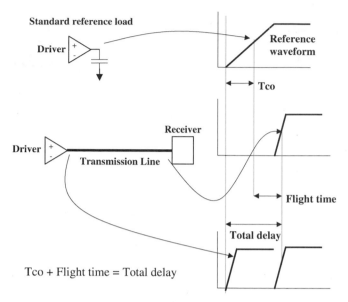

FIGURE 9.6 Proper use of a reference load in calculating total delay.

into a standard reference load. The middle waveform depicts the waveform at the receiver. The bottom waveform depicts the waveforms at the driver loaded by the system and its timing relationship to the receiver. Notice that if the standard reference load is used to calculated both the flight time and the T_{co}, when they are summed in the spreadsheet the total delay, which is what we care about, comes out correctly.

This leads us to a few conclusions. First, the flight time is *not* the same as the propagation delay (flight time is explained in more detail in Section 9.2.3) and the T_{co} as measured into the standard reference load is not the same as it will be in the system. The total delay, however, is the same. Note that if the reference load is chosen correctly, the flight times will be very similar or identical to the propagation delays, and the T_{co} timings will be very similar to what the system experiences.

The reference load for CMOS buffers are often just capacitors that represent the input capacitance seen by the loaded output buffer (if a transmission line was used, there would be a ledge in the reference waveform, which would cause difficulties when calculating timings). This capacitance typically includes the distributed capacitance seen over the duration of the signal edge. For example, if an output buffer with an edge speed of 300 ps is driving a transmission line with a C_{11} of 2.5 pF/in. and a propagation delay of 150 ps/in., the edge would encompass 2 in. or 5 pF. Subsequently, a good load in this case would be 5 pF.

The reference load for a GTL bus usually looks more like the system input because GTL buffers do not exhibit ledges at the drivers as do CMOS buffers

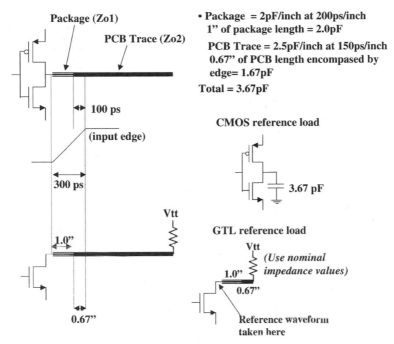

FIGURE 9.7 Examples of good reference loads.

when loaded with a transmission line (see Appendix B for current-mode analysis of GTL buffers). Figure 9.7 depicts an example of reference loads for a GTL and a CMOS output buffer. The CMOS reference load is simply a capacitor that is equivalent to the system input capacitance seen for the duration of the edge. This will exhibit sufficient signal integrity for a reference load and it will load the buffer similar to the system. The GTL buffer resembles the input of the system. In this particular case, the system is duplicated for the duration of the edge, and the pull-up is included because it is necessary for proper operation. In both cases, typical values should be used for the simulation reference load. It is not necessary to account for process variations when developing the reference load. The simple rule of thumb is to observe the system input for the duration of the edge and use it to design the reference load. Make certain that the signal integrity of the reference load exhibits no ledges and minimal ringing.

It should be noted that this same technique is used to insert timings from a high-volume manufacturing tester into a design validation spreadsheet when the final system has been built and is being tested in the laboratory. The difference is that a standard tester load is used instead of a standard simulation load. However, since it may be impossible to use identical loads for silicon- and system-level testing, simulation can be used to correlate the timings so they add up correctly in the design validation spreadsheet.

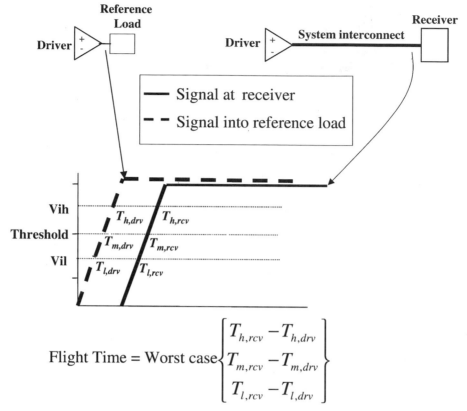

$$\text{Flight Time} = \text{Worst case} \begin{cases} T_{h,rcv} - T_{h,drv} \\ T_{m,rcv} - T_{m,drv} \\ T_{l,rcv} - T_{l,drv} \end{cases}$$

FIGURE 9.8   Standard flight-time calculation.

9.2.3. Flight Time

Flight time is very similar to propagation time, which is the amount of time it takes a signal to propagate along the length of a transmission line, as defined in Chapter 2. Propagation delay and flight time, however, are not the same. *Propagation delay* refers to the absolute delay of a transmission line. *Flight time* refers to the difference between the reference waveform and the waveform at the receiver (see Section 9.2.2). The difference is subtle, but the calculation methods are quite different because a simulation reference load is required to calculate flight time properly so that it can be inserted into the timing spreadsheets.

Flight time is defined as the amount of time that elapses between the point where the signal on an output buffer driving a reference load crosses the threshold voltage and the time where the signal crosses the threshold voltage at the receiver in the system. Figure 9.8 depicts the standard definition of flight time. The flight time should be evaluated at V_{il}, $V_{threshold}$, and V_{ih}. The methodology is to measure the flight time at these three points and insert the applicable worst case into the spreadsheet. It should be noted that sometimes the worst-case value is the maximum flight time, and other times it is the minimum flight time. For example, the maximum flight time produces the worst-case condition in a common-clock setup margin calculation, and the minimum flight time produces the worst case for the hold margin. To ensure that the worst-case condition is captured, the timing equations should be observed.

A common mistake is to measure flight time from V_{il} on the reference driver to V_{ih} at the receiver (or from V_{ih} to V_{il}). This is not correct because it counts the time it takes for the signal to transition through the threshold region as part of the delay. This delay is already accounted for in the T_{co} variable of the timing equations.

9.2.4. Flight-Time Skew

Flight-time skew is simply the difference in flight times between two nets. In a source synchronous design, flight-time skew is defined as

$$T_{\text{flight skew}} = T_{\text{flight data}} - T_{\text{flight strobe}} \tag{9.5}$$

It is standard terminology to define skew as data flight time minus strobe flight time. Figure 9.9 depicts a generic example of flight-time skew calculation for a GTL bus.

If the timing equations in Chapter 8 are studied, it is evident that the worst-case condition for hold-time margin is when the data arrive early and the strobe arrives late. The following equation calculates the minimum skew for use in the source synchronous hold equation:

$$T_{\text{skew, min}} = (T_{\text{flight data}})_{\min} - (T_{\text{flight strobe}})_{\max} \tag{9.6}$$

Note that this equation is not sufficient to account for the skew in a real system. Although in the preceding section we warned against calculating flight time across the threshold region, it is necessary to calculate skew (between data and strobe) across the threshold region for conventional source synchronous designs. As described in Chapter 8, the data and the strobe can be separated by several nanoseconds assuming a conventional source synchronous design. During this time, noise may be generated in the system (i.e., by core operations, by multiple outputs switching simultaneously, etc.) and the threshold voltage may change between data and strobe. Subsequently, it is necessary to calculate skew by subtracting flight times at different thresholds to yield the worst-case results. To do so, the skew is calculated at the upper and lower boundaries of the threshold region. The following equation calculates the minimum skew while accounting for the possibility of noise causing variations of

Reference load for strobe   Reference load for data

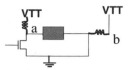

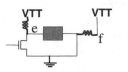

Package & socket models

Strobe net   Data net

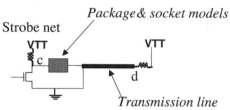

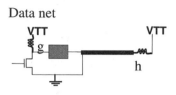

Transmission line

Flight time calculation of Flight skew calculation between data & strobe
data net & strobe

$$\text{Skew} = \text{FT(data)} - \text{FT(strobe)} = T_h - T_e - T_d + T_a$$

FT(strobe) = T_d-T_a
FT(data) = T_h-T_e

If the strobe and data buffers
are identical: Skew = T_h-T_d

FIGURE 9.9 Simple example of flight-time skew calculation. Note that the load is identical for data and strobe, but the driving buffer may differ.

the threshold voltage with respect to the signal:

$$T_{\text{skew, hold}} = \min \left\{ \begin{array}{c} (T_{h,\text{rcv}})_{\text{data}} - (T_{h,\text{drv}})_{\text{data}} \\ (T_{l,\text{rcv}})_{\text{data}} - (T_{l,\text{drv}})_{\text{data}} \end{array} \right\} - \max \left\{ \begin{array}{c} (T_{h,\text{rcv}})_{\text{strobe}} - (T_{h,\text{drv}})_{\text{strobe}} \\ (T_{l,\text{rcv}})_{\text{strobe}} - (T_{l,\text{drv}})_{\text{strobe}} \end{array} \right\}$$

$$(9.7)$$

where $T_{h,\text{rcv}}$ is the time at which the signal at the receiver crosses V_{ih}, $T_{h,\text{drv}}$ the time at which the signal at the input of the reference load crosses V_{ih}, $T_{l,\text{rcv}}$ the time at which the signal at the receiver crosses V_{il}, and $T_{l,\text{drv}}$ the time at which the signal at the input of the reference load crosses V_{il}. Note that even though the skew may be calculated across the threshold region, the flight times are not.

The worst-case condition for source synchronous setup margin is when the data arrive late and the strobe arrives early. The maximum skew for use in the source synchronous setup equations is calculated as

$$T_{\text{skew, max}} = (T_{\text{flight data}})_{\text{max}} - (T_{\text{flight strobe}})_{\text{min}} \qquad (9.8)$$

$$T_{\text{skew, setup}} = \max \left\{ \begin{array}{c} (T_{h,\text{rcv}})_{\text{data}} - (T_{h,\text{drv}})_{\text{data}} \\ (T_{l,\text{rcv}})_{\text{data}} - (T_{l,\text{drv}})_{\text{data}} \end{array} \right\} - \min \left\{ \begin{array}{c} (T_{h,\text{rcv}})_{\text{strobe}} - (T_{h,\text{drv}})_{\text{strobe}} \\ (T_{l,\text{rcv}})_{\text{strobe}} - (T_{l,\text{drv}})_{\text{strobe}} \end{array} \right\}$$

$$(9.9)$$

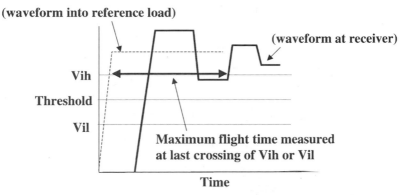

FIGURE 9.10   Flight-time calculation when the signal rings back into the threshold region.

9.2.5. Signal Integrity

Signal integrity, sometimes referred to as *signal quality*, is a measure of the waveform shape. In a digital system, a signal with ideal signal integrity would resemble a clean trapezoid. A signal integrity violation refers to a waveform distortion severe enough either to violate the timing requirements of the system or to blow out the gate oxides of the buffers. In the following section we define the basic signal integrity metrics used in the design of a high-speed digital system.

Ringback. *Ringback* is defined as the amount of voltage that "rings" back toward the threshold voltage. Figure 9.10 graphically depicts a ringback violation into the threshold region as measured at a receiver. The primary effect of a ringback violation into the threshold region is an increase in the flight time. If the signal is ringing and it oscillates into the threshold region, the maximum flight time is calculated at the point where it last leaves the threshold region, as depicted in Figure 9.10. This is because the buffer is in an indeterminate state (high or low) within the threshold region and cannot be considered high or low until it leaves the region. This can also dramatically increase the skew between two signals if one signal rings back into the threshold region and the other signal does not.

It is vitally important that strobe or clock signals not ring back into the threshold region because each crossing into or out of the region can cause false triggers. If the strobe in a source synchronous bus triggers falsely, the wrong data will be latched into the receiver circuitry. However, if a strobe that runs a state machine triggers falsely, a catastrophic system failure could result. Ringback is detrimental even if it does not cross into the threshold region. If a signal is not fully settled before the next transition, ISI is exacerbated (see Chapter 4).

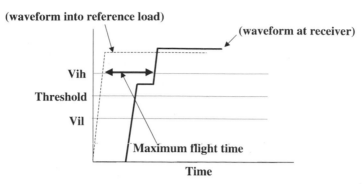

FIGURE 9.11   Flight-time calculation when the edge is not linear through the threshold region.

Nonmonotonic (Nonlinear) Edges. A nonmonotonic edge occurs when the signal edge deviates significantly from linear behavior through the threshold region, such as temporarily reversing slope. This is caused by several effects, such as the stubs on middle agents of a multidrop bus, as in Figure 5.16; return path discontinuities, as in Figure 6.3; or power delivery problems, as in Figure 6.12. The primary effect is an increase in the effective flight time or in the flight-time skew if one signal exhibits a ledge and the other does not. Figure 9.11 shows the effective increase in the signal flight time as measured at V_{ih} for a rising edge.

Overshoot/Undershoot. The maximum tolerable overshoot and undershoot are a function of the specific circuitry used in the system. Excessive overshoot can break through the gate oxide on the receiver or driver circuitry and cause severe product reliability problems. Sometimes the input buffers have diode clamps to help protect against overshoot and undershoot; however, as system speeds increase, it is becoming increasingly difficult to design clamps that are fast enough to prevent the problem. The best way to prevent reliability problems due to overshoot/undershoot violation is with a sound interconnect design. Overshoot/undershoot does not affect flight time or skews directly, but an excessive amount can significantly affect timings by exacerbating ISI. Figure 9.12 provides a graphic definition of overshoot/undershoot.

9.3. DESIGN OPTIMIZATION

Now that the initial timings and the signal quality/timings metrics have been defined, it is time to begin the actual design. First, let's discuss the old method of bus design. Figure 9.13 is a flowchart that shows the wrong way to design a high-speed bus. This methodology requires the engineer to route a portion of the board, extract and simulate each net on the board, fix any problems, and reroute. As bus speeds increase and the number of variables expand, this old

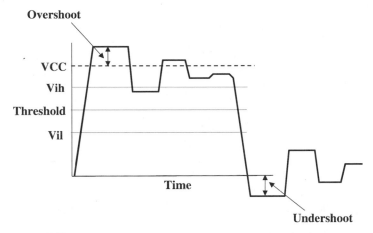

FIGURE 9.12 Definition of overshoot and undershoot.

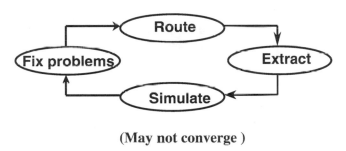

(May not converge)

FIGURE 9.13 Wrong way to optimize a bus design.

methodology becomes less efficient. For multiprocessor systems, it is almost impossible to converge using this technique.

A more efficient method of design process requires a structured procedure to ensure that all variables are accounted for in the pre-layout design. If this is done correctly, a working layout will be produced on the first pass and the extraction and simulation process is used only to double check the design and ensure that the layout guidelines were adhered to. Figure 9.14 depicts a more efficient methodology for high-speed bus design.

9.3.1. Paper Analysis

After the initial timings and the metrics have been defined, the next step is to determine all the signal categories contained within the design. At this point, the engineer should identify which signals are source-synchronous, common-clock, controls, clocks, or fall into another special category. The next portion of the paper analysis is to determine the best estimates of all the system variables. The estimates should include mean values and estimated maximum variations.

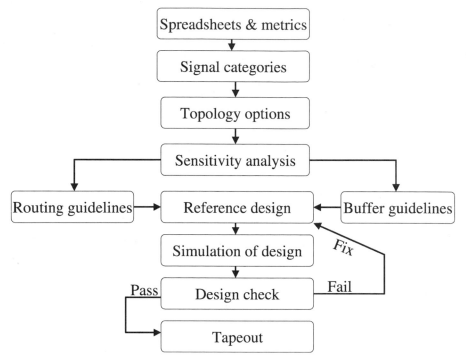

FIGURE 9.14   Efficient bus design methodology.

It is important to note that these are only the initial rough estimates and all are subject to change. This is an important step, however, because it provides a starting point for the analysis. Each of the variables will be evaluated during the sensitivity analysis and changed or limited accordingly. Some examples of system variables are listed below.

- I/O capacitance
- Trace length, velocity, and impedance
- Interlayer impedance variations
- Buffer strengths and edge rates
- Termination values
- Receiver setup and hold times
- Interconnect skew specifications
- Package, daughtercard, and socket parameters

9.3.2. Routing Study

Once the signals have been categorized and the initial timings have been determined, a list of potential interconnect topologies for each signal group must be determined. This requires significant collaboration with the layout engineer and is developed through a layout study. The layout and design engineers should work together to determine the optimum part placement, part pin-out, and all possible physical interconnect solutions. The layout study will produce a *layout solution space*, which lists all the possible interconnect topology options, including line lengths, widths, and spacing. Extensive simulations during the sensitivity analysis will be used to limit the layout solution space and produce a final solution that will meet all timing and signal quality specifications. During the sensitivity analysis, each of these topologies would be simulated and compared. The best solution would be implemented in the final design.

The layout engineer should collaborate with the silicon and package designers very early in the design cycle. Significant collaboration should be maintained throughout the design process to ensure adequate package pin-out and package design. It is important to note that the package pin-out could "make or break" the system design.

Topology Options. During the layout study it is beneficial to involve the layout engineers in determining which topologies are adequate. Figure 9.15 depicts several different layout topologies that are common in designs. It is important to note, however, that the key to any acceptable routing topology is symmetry (see Chapter 4).

1. *Point to point.* This is the simplest layout option. The major routing restrictions to worry about are minimum and maximum line lengths and the ability to match the line lengths in signal groups.

2. *Heavy point to point.* This topology has the same restrictions as above. Furthermore, the short stubs that connect the main interconnect to the components should be kept very short; otherwise, the topology will behave like a star or T-topology.

3. *T topology.* T topologies are usually unidirectional; that is, agent 1 (from Figure 9.15*d*) is the only driver. As mentioned above, symmetry is necessary for adequate signal integrity. The T topology is balanced from the view of agent 1 as long as the legs of the T are the same lengths. Assuming that the legs of the T are equal in length, the driver will only see an impedance discontinuity equal to one-half the characteristic impedance of the transmission lines. If the T is not balanced (such as when agent 2 or 3 is driving), it will resonate and the signal integrity will be severely diminished. If all three legs were made to be equal, the bus would become unidirectional because it would be symmetrical from all agents (then the topology would be a three-agent star, not a T). One trick that is sometimes used to improve T topologies is to make the T legs twice the impedance of the base, which will eliminate the impedance discontinuity at the junction (see Section 4.6).

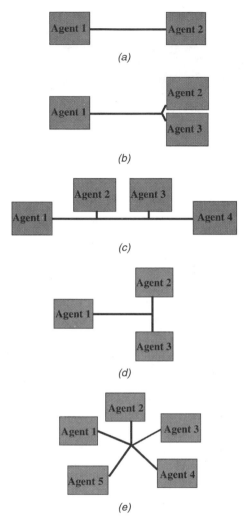

FIGURE 9.15 Common bus topologies: (*a*) point to point; (*b*) heavy point to point; (*c*) daisy chain; (*d*) T topology; and (*e*) star topology.

4. *Star topology*. When routing a star topology, it is critical that the electrical delay of each leg be identical; otherwise, the signal integrity will degrade quickly. Star topologies are inherently unstable. It is also important to load each leg the same to avoid signal integrity problems.

5. *Daisy chain*. This is a common topology for multidrop buses such as front-side buses and memory buses on personal computers. The major disadvantage is the fact that stubs are usually necessary to connect the middle agents to the main bus trunk. This causes signal integrity problems. Even when the stubs are extremely short, the input capacitance of

the middle agents load the bus and can lower the effective characteristic impedance of the line as described in Section 5.3.3.

Early Consideration of Thermal Problems. After the layout solution space has been determined, it is a good idea to consider the thermal solution or consult with the engineers who are designing the thermal solution. Often, the interconnect design in computer systems are severely limited by the thermal solution. Processors tend to dissipate a significant amount of heat. Subsequently, large heat sinks and thermal shadowing limit minimum line lengths and component placing. If the thermal solution is considered early in the design process, severe problems can be avoided in the future.

9.4. SENSITIVITY ANALYSIS

The sensitivity analysis constitutes the bulk of the design. The routing study is used to determine the layout solution space. The sensitivity analysis is used to determine the *electrical solution space*. The area where both of these solutions spaces overlap constitutes the *design solution space*. The sensitivity analysis will rank the variables as to how strongly they affect the system and highlight which variables affect the solution most and in what way they affect the system. This allows the designer to get a hold of a problem with a great number of variables.

During a sensitivity analysis, every variable in the system bus is varied in simulation. The performance metrics, such as flight time, flight-time skew, and signal integrity, are observed while each variable is swept. The performance as a function of each variable is compared to the timing and signal quality specifications. The result is a solution space that will place strict limits on the system variables under control of the designer (such as trace lengths, spacing, impedance, etc.). The solution space will lead to design guidelines for the PCB, package, connectors, termination, and I/O.

9.4.1. Initial Trend and Significance Analysis

An initial Monte Carlo (IMC) analysis can be used to determine the significance and trend of each variable in the system. This is necessary for two reasons: First, the trends will point to the worst- and best-case conditions. This will allow the engineer to define the simulations at the worst- and best-case corners. In a design with a large number of variables, it is almost impossible to pick the worst-case conditions by reasoning alone. The IMC analysis will point directly to the corner conditions. Second, the IMC can be used to determine which variables most affect performance. Subsequently, significant time can be saved by prioritizing the analysis and concentrating primarily on the most significant variables.

After a rough topology and all the system variables have been estimated, a Monte Carlo analysis should be performed. Every single variable in the system should be varied with uniform distribution over a reasonable range that approximates the manufacturing variations. It is not necessary to perform more than 1000 iterations because the analysis is not meant to be exhaustive. After the IMC is performed, each variable is plotted versus a performance metric such as flight time, skew, or ringback. A linear curve is then fit to the data so that the trend and correlation can be observed. A linear relationship is usually adequate because all the input variables at this stage should be linear. Furthermore, since these variables are being swept, it is usually optimal to use linear buffers. So, basically, what is that done is that a great number of variables are allowed to vary randomly (Monte Carlo), and the results for several defined timing and signal integrity metrics are recorded (usually automatically from scripts written to filter the output or utilities contained in the simulation software). The observed metrics can then be plotted versus system variables. If the metric is strongly dependent on the system variable, it will "stand out" of the blur and be easily recognizable.

Figure 9.16 depicts a conceptual IMC output. The plot depicts the flight time from a Monte Carlo analysis (with several variables) plotted versus only one variable, the receiver capacitance. A best linear fit was performed on the data and the correlation factor was calculated using Microsoft Excel. The correlation factor is a measure of how well the linear approximation represents the data. A value of 1.0 indicates that the linear approximation is a perfect representation of the data. A value of 0.0 indicates that no correlation exists between the linear approximation and the data. In an IMC analysis, the correlation factor can be used to determine how dependent the overall metric is on the particular variable being plotted (in this case the flight time versus capacitance). If the correlation factor is high, the metric is highly dependent on the variable. If the metric has no linear trend when plotted against the variable, the correlation factor will be very small. In the case of Figure 9.16, since the correlation factor is 0.96, out of all the parameters varied in the Monte Carlo analysis, the flight time is almost exclusively dependent on the I/O capacitance. The slope of the linear fit indicates the trend of the variable

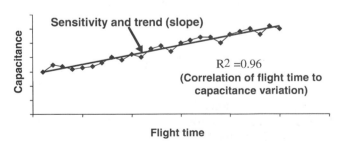

FIGURE 9.16   Receiver capacitance versus flight time: IMC analysis that shows maximum receiver capacitance produces the longest flight time.

versus the metric. The positive slope indicates that the worst-case flight time will be achieved with large capacitance values. Furthermore, if it is difficult to obtain the correlation factor, the magnitude of the slope provides insight into the significance of the variable. If the slope is flat, it indicates that the variable has minimal influence on the metric. If the slope is large, the variable has a significant influence on the variable. In large complex systems, this data is invaluable because it is sometimes almost impossible to determine the worst-case combinations of the variables for timings and signal integrity without this kind of analysis.

It should be noted that sometimes a small number of variables will dominate a performance metric. For example, length and dielectric constant will dominate flight times. When these variables are included in a large IMC analysis, the correlation factors associated with them will be very high and all the others will be very low. Subsequently, all the trend and significance data associated with the other variables will be pushed into the noise of the simulation. Usually, after the initial IMC analysis, it is necessary to limit the variables for subsequent runs by fixing the dominant variables. This will pull the significance and trend data out of the noise for the less dominant variables.

Figure 9.16 was an ideal case. For a more realistic example, refer to Figure 9.17. A Monte Carlo analysis was performed on a three-load system and every variable in the system was varied randomly with a uniform distribution over reasonable manufacturing tolerances (19 variables in this case). Figure 9.17

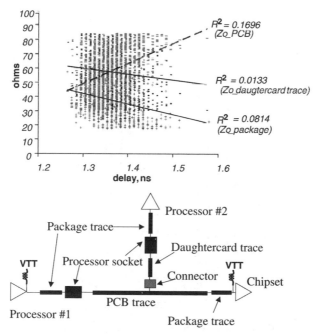

FIGURE 9.17 Using IMC analysis to determine impedance significance and trends.

is a scatter plot of the delays produced during a 1000-iteration Monte Carlo analysis plotted against the impedance of the motherboard, the daughtercard trace, and the package trace. The best linear fit was performed on each data set. Both the trends and the significance are determined by the results of the linear approximation. Since the slope of the line is positive for the PCB impedance, the delay of the nets will increase as the trace impedance increases. Furthermore, since the slopes of the package and daughtercard impedance are negative, this means that the longest delay for this particular system will be achieved when the PCB impedance is high and the package and daughtercard impedance are low. This result is definitely not intuitive; however, it can be verified with separate simulations. Additionally, the correlation factor is used to determine the significance of the variables. Since the correlation factor of the PCB impedance is the highest, it has more influence on the signal delay than does the daughtercard or package impedance.

Figure 9.18 is a plot of the same Monte Carlo results as those shown in Figure 9.17 plotted versus the I/O capacitance and the socket. It is easy to see the trend of the delay with increasing capacitance values. Plots such as this should be made for every variable in the system. For example, these two plots indicate that the worst-case delay will happen when the I/O capacitance and the PCB impedance is high and the package and daughtercard impedance are low. Similar plots can be made to determine the worst-case conditions for skew, overshoot, and ringback.

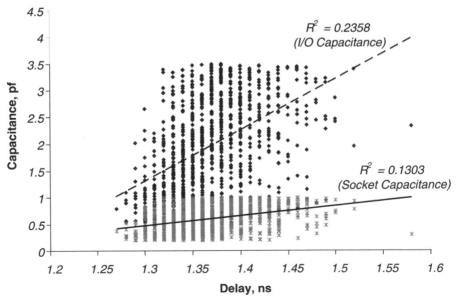

FIGURE 9.18 Using IMC analysis to determine the significance and trends of I/O and socket capacitance.

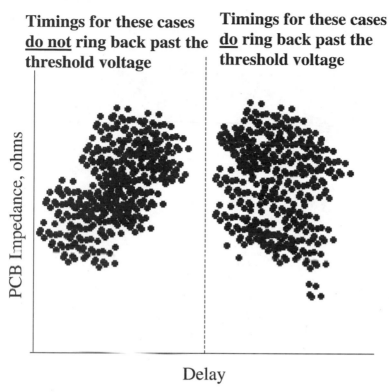

Timings for these cases do not ring back past the threshold voltage

Timings for these cases do ring back past the threshold voltage

Delay

FIGURE 9.19 Output of an IMC analysis when the system exhibits ringback errors.

It should be noted these trends are feasible only when the system is well behaved. Figure 9.19 is an IMC analyses for an ill-behaved system. Notice the distinctive gap in the scatter plot. This is because ringback becomes large enough that some of the iterations violate the switching threshold. Subsequently, the point where the ringback last crosses the threshold is considered the flight time. Figure 9.20 is a representative waveform with a ringback violation taken from the Monte Carlo simulation. The points less than 1 ns are the delay measured at the rising edge when ringback does not violate the threshold region, and the points greater than 1 ns are the delay measured at the ringback violation when it occurs. If such patterns occur during a IMC analysis, the data should be analyzed to determine which variables are causing the system to be ill behaved. Steps should then be taken to confine those variables and the IMC should be repeated. As with all analysis, the designer must view the waveforms to ensure that the results are being interpreted correctly.

After this initial analysis has been performed and all significance and trends have been analyzed, the worst-case corners should be identified and the variables should be prioritized. The worst-case conditions (also known as *corner*

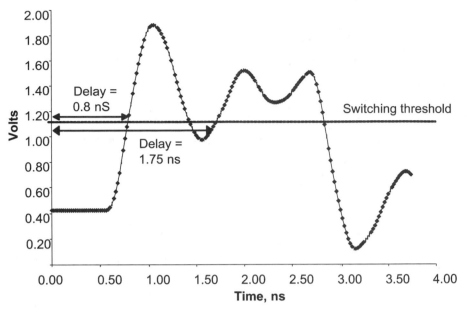

FIGURE 9.20 Waveform from the IMC analysis depicted in Figure 9.19 that shows how ringback violations can cause a gap in the scatter plot.

conditions) should be identified for the following eight cases:

1. Maximum flight time
2. Minimum flight time
3. Maximum high-transition ringback
4. Maximum low-transition ringback
5. Maximum overshoot
6. Maximum undershoot
7. Maximum flight-time skew
8. Minimum flight-time skew

Furthermore, a list of the variables in decreasing order of significance (based on the linear fits described previously) should be created for each of the worst-case corners (known as a *significance list*). The significance list can help to prioritize the analysis and troubleshoot later in the design.

The maximum and minimum flight-time skews can be surmised from the flight-time plots because flight-time skew is simply the difference in flight times. The worst-case conditions should be determined assuming maximum variations possible for a given system. For example, the total variation over all processes for buffer impedance may vary 20% from part to part. However, impedance may not vary more than 5% for buffers on a single part. Subsequently, if the maximum flight time skew occurs when the output impedance for the data net is high and the strobe impedance is low, the worst-case conditions should reflect only the maximum variation on a single part, not the entire process range for all parts (assuming that data and strobe are generated from the same part). In other words, when defining the worst-case conditions, be certain that you remain within the bounds of reality.

9.4.2. Ordered Parameter Sweeps

After the IMC analysis is complete and the eight corner conditions are determined, the solution space needs to be resolved. The solution space puts limits on all the variables that the designer has control of so that the system timing and signal integrity specifications are not violated. The ordered parameter sweeps are used to sweep one or two variables at a time systematically so that the behavior of the system can be understood and parameter limits can be chosen. All the fixed variables (the ones that are not being swept) should be set at the worst-case (corner) conditions during the sweeps. The worst-case conditions should be determined from the IMC analysis. Figure 9.21 is an example of a three-dimensional parameter sweep (the topology is shown in Figure 9.17) used to determine the maximum flight time. The fixed variables such as the I/O capacitance and the stub lengths were held constant at the worst-case values for maximum flight time while the motherboard and daughtercard impedance were swept in value. The values of this sweep are compared to the specifications. In this case it can readily be seen that the

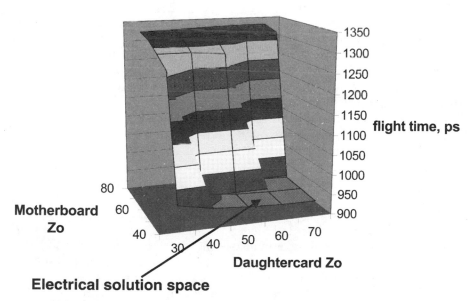

FIGURE 9.21 Example of a three-dimensional ordered parameter sweep.

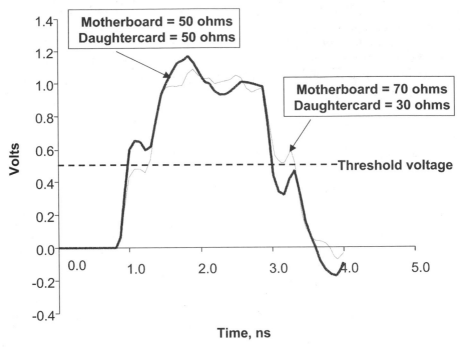

FIGURE 9.22 Waveforms taken from the simulation that created the parameter sweep in Figure 9.21. Note the sudden increase in flight time when the ledge moves below the threshold region.

flight time increased dramatically when the motherboard impedance is greater than 60 Ω. The middle stub (daughtercard trace) causes the sudden increase in skew. As described in Section 5.3.3, as the impedance of the stub is decreased relative to the main bus trace impedance, the ledge moves down. The plot in Figure 9.21 is flight time taken at a threshold voltage of $V_{cc/2}$. The sharp jump in flight time occurs when the ledge moves below the threshold voltage. Figure 9.22 depicts waveforms at two different locations of the three-dimensional plot to show how the position of the ledge produces the three-dimensional shape. It is readily seen that the combination of the low stub impedance and the high board impedance pushed the ledge below the threshold voltage, causing an increase in the flight time equal to the length of the ledge.

Plots such as in Figures 9.21 and 9.23 are used to limit the solution space. For example, the plot shown in Figure 9.23, which is simply the top view of Figure 9.21, can be used to determine the impedance boundaries for both the motherboard and daughtercard trace so that the maximum flight time is not exceeded. In this case, assuming a maximum tolerable flight time of 1000 ps, the total impedance variation, including manufacturing variations and crosstalk, must be contained within the dashed line.

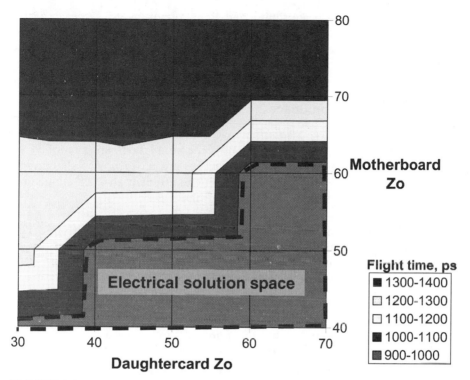

FIGURE 9.23 Top view of Figure 9.21. This view is useful in defining the solution space.

The example shown in Figure 9.23 will produce two solutions spaces:

1. $38\ \Omega <$ daughtercard $Z_o < 70\ \Omega$ and $40\ \Omega <$ motherboard $Z_o < 52\ \Omega$
2. $58\ \Omega <$ daughtercard $Z_o < 70\ \Omega$ and $40\ \Omega <$ motherboard $Z_o < 61\ \Omega$

Furthermore, it can be used to estimate the minimum spacing. Since the impedance is varied over a wide area, the maximum tolerable impedance variations can be translated into minimum spacing requirements by reverse engineering a SLEM model (the SLEM model is described in Section 3.6.2). A SLEM model translates crosstalk, which is a function of line–line spacing into an impedance variation. It is not difficult to translate an impedance variation into a spacing requirement for a given cross section. *Note*: Be certain that the manufacture impedance variations are accounted for when translating an impedance variation into a spacing requirement.

It is not necessary to use three-dimensional plots; however, they are generally more efficient than a two-dimensional approach. The parameter sweeps are then used to limit each variable that the designer has control of in order to choose the optimal electrical solution space. It is necessary to perform parametric sweeps for all relevant timing and signal quality metrics at each of the eight corners defined in Section 9.4.1. Often, custom scripts must be written to automate these sweeps efficiently. However, many simulation tool vendors are beginning to offer these capabilities.

9.4.3. Phase 1 Solution Space

The purpose of the phase 1 solution space is to limit the electrical solution space from infinite to finite. Subsequently, the simulations are simplified so that they can run fast. Ordered parameter sweeps are performed on all the variables as described in Section 9.4.2. To facilitate efficient and fast simulation times, several assumptions are made during phase 1.

1. SLEM transmission line models are used to account for crosstalk. The impedance is swept and the allowable impedance range is translated into an initial target impedance and spacing requirement. It is important to allow for manufacturer variations.
2. Power delivery and return paths are assumed to be perfect.
3. Simple patterns are used (no long pseudorandom patterns).
4. Simple linear buffers are used that are easy to sweep.

With these assumptions in place, the designer can quickly run numerous simulations to pick the topologies, the initial line lengths, minimum trace spacing, buffer impedances, termination values and PCB impedance values. Future simulations will incorporate the more computationally intensive effects. After the completion of the phase 1 solution space, the initial buffer parameters, such as impedance and edge rates, should be given to the silicon designers. This will allow them to design the initial buffers, which are needed for the phase 2 solution space.

Phase 1 ISI. Intersymbol interference (ISI) also needs to be accounted for in the parameter sweeps. ISI is simply pattern-dependent noise or timing variations that occur when the noise on the bus has not settled prior to the next transition (see Section 4.4). To estimate ISI in a system, it is necessary to simulate long pseudorandom pulse trains. This technique, however, is long, cumbersome, and very difficult to do during a parameter sweep. The following technique will allow the designer to gain a first-order approximation of the ISI impact in the phase 1 solution space. The full effect of the ISI will be evaluated later.

To capture most of the timing impacts due to intersymbol interference, the parameter sweeps (for all eight worst-case corners) should be performed at the fastest bus frequency, and then at $2\times$ and $3\times$ multiples of the fastest bus period. For example, if the fastest frequency the bus will operate at produces a single bit pulse duration of 1.25 ns, the data pattern should be repeated with pulse durations of 2.5 and 3.75 ns. This will represent the following data patterns transitioning at the highest bus rate.

010101010101010

001100110011001

000111000111000

The timings should be taken at each transition for at least five periods so that the signal can reach steady state. The worst-case results of these should be used to produce the phase 1 solution space. This will produce a first-order approximation of the pattern-dependent impact on the bus design. This analysis can be completed in a fraction of the time that it takes to perform a similar analysis using long pseudorandom patterns. Additionally, since the worst signal integrity will often occur at a frequency other than the maximum bus speed, this technique helps ensure that signal integrity violations are not masked by the switching rate as in Figure 4.16.

Monte Carlo Double Check. After the phase 1 solution space is determined, it is a good idea to perform a final check. Monte Carlo analysis can used as a double check on the phase 1 solution space to ensure passing margin under all conditions. Although the IMC analysis is designed to ensure that the parameter sweeps will yield the worst-case conditions, it is possible that the worst-case combination of variables was not achieved. During the Monte Carlo analysis, the results of the metric (i.e., the flight time or the skew) should be observed, as all the individual components of the design are bounded by the constraints of the phase 1 solution space and varied randomly. The maximum and minimum values should be compared to the specifications. If there are any violations, the specific conditions that caused the failure should be observed so that the mechanism can be established. Therefore, it is necessary to keep track of the Monte Carlo output *and* the random input variables so that a specific output can be correlated to all the input variables. It is sometimes useful to plot the output of the Monte Carlo analysis against specific variables as was done in the initial Monte Carlo analysis. This often provides significant insight into the cause of the failure.

Note that using Monte Carlo analysis as a double check will not necessarily capture any mistakes, so do not rely on it. Monte Carlo at this stage is simply a "shotgun blast in the dark"—maybe you will hit something and maybe you won't. If there is a mistake in your solution space, hopefully you will find it.

9.4.4. Phase 2 Solution Space

The phase 2 solution space in an incremental step toward the final solution space. In the phase 1 solution space, many approximations were made to quickly limit the major variables of the solution space. In the development of the phase 2 solution space, a limited set of simulations are performed near the edges of the phase 1 solution space. These simulations should contain fully coupled transmission lines instead of a SLEM approximation. IBIS or transistor models should be used in place of the linear buffer models, and full ISI analysis with pseudorandom patterns should be used.

Phase 2 solution space augments the phase 1 solution space by removing some of the approximations as listed below.

1. The SLEM models should be replaced by fully coupled transmission line models with the minimum line–line spacing. Remember that the minimum spacing was determined from the impedance sweeps in the phase 1 solution space by observing maximum and minimum impedance variations. A minimum of three lines should be simulated, with the middle line being the target net.

2. A long pseudorandom bit pattern should be used in the simulation to determine the impact of ISI more accurately. This pattern should be chosen to maximize signal integrity distortions by switching between fast and slow data rates and between odd- and even-mode switching patterns.

3. The linear buffers should be replaced with full transistor or IBIS behavioral models.

The phase 2 solution space will further limit the phase 1 solution space. Again, after the completion of phase 2, it is a good idea to perform a large Monte Carlo simulation to double check the design space.

Eye Diagrams Versus Parameter Plots. During the phase 1 analysis, parameter plots are used to limit the large number of variables in the system. Since many approximations are made, it is relatively simple to create a UNIX script or write a C++ program to sort though the output data from the simulator and extract the timings from the waveforms. Furthermore, some interconnect simulation suites already have this capability built in. Parameter plots are very efficient when running extremely large sets of simulations because they are faster and they consume significantly less hard-drive space that an eye diagram sweep. Furthermore, parameter plots give the designer an instant understanding of the trends.

An eye diagram is created by laying individual periods of transitions on a net on top of each other. An eye diagram is useful for analyzing individual, more complicated simulations (not sweeps) with randomized data patterns. The main advantage of an eye diagram is that it is more reminiscent of a laboratory measurement, and several useful data points can be obtained simply by looking at the eye. The disadvantage is that they are cumbersome, not well suited to the analysis of many simulations, and do not provide an easy way to separate out silicon- and system-level effects.

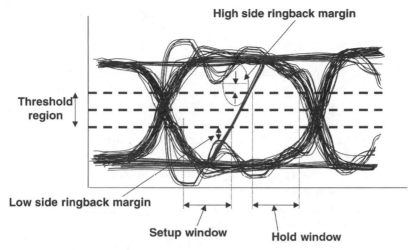

FIGURE 9.24   Example of an eye diagram.

Figure 9.24 depicts an example of an eye diagram from a phase 2 source synchronous simulation. The strobe is superimposed on top of the data eye. Several effects are easily identifiable. The total setup and hold windows are determined simply by measuring the time between the strobe and data at the upper or lower portion of the noise margin. If the setup/hold windows are greater than the receivers setup/hold requirements, then the timing margin is positive. Furthermore, the ringback margin can be determined simply by observing how close the signal rings back toward the threshold voltage. It should be obvious that eye diagrams are not particularly suited for sweeping, or for evaluation of flight time or flight-time skews. However, many engineers prefer to utilize them in the later stages of the design.

Phase 2 ISI. The evaluation of ISI can be tricky. To do so, it is necessary to perform targeted simulations at the edge of the design space using a long pseudorandom pulse train. If pattern-dependent timing or signal quality violations occur, steps should be taken to minimize reflections on the bus; this will reduce the ISI. Usually, the best way to limit reflections on the bus is to tighten up the impedance variations and minimize discontinuities by shortening stubs and connectors and matching impedances between packages and PCBs. The results of the ordered sweeps and the significance list should be quite helpful at this stage. Also, be sure to include any startup strobe ISI in the analysis. Usually, strobes do not exhibit ISI because they operate in a single pattern. However, often the strobe is turned off until it is time for the component with which it is associated to drive signals onto the bus. It will take several transitions for the strobe to reach steady state. If the bus protocol allows valid signals to be driven onto the bus before the strobe can reach steady state, the strobe ISI can significantly affect the timings by increasing skew.

The worst-case ISI can be evaluated using the following steps:

1. Simulate the longest net in the bus with the most impedance discontinuities using a long pseudorandom bit pattern for each of the worst-case conditions (i.e., they should be at the edge of the solution space so far defined).
2. Take the flight time of the first transition of the ISI simulation as a baseline.
3. Determine the rising and falling flight times for each bus transition (using the standard load as a reference).
4. Subtract the minimum and maximum delays from the baseline delays and find the worst-case difference.
5. Take the smallest negative and the greatest positive difference. This should be the worst-case ISI impact on timings.
6. Simulate the maximum and minimum flight time and flight-time skew corners with a very slow switching rate that settles completely prior to each transition and add the numbers from step 5 to the results. This should represent the worst-case phase 2 flight time and flight-time skew numbers that should be included in the spreadsheet. Remember to account for the startup ISI on the strobe if necessary.
7. Look for the worst-case overshoot, undershoot, and ringback, and make sure that they do not violate the threshold region or exceed maximum overshoot specifications.

The alternative method is to use an eye diagram. Note that if an eye diagram is used, the total timings measured will already incorporate ISI.

9.4.5. Phase 3 Solution Space

The phase 1 and phase 2 solution spaces have significantly narrowed the solution space. Targets for all aspects of the design should be defined. The phase 3 solution space is the final step. Phase 3 should include the effects that are very difficult to model. The timing impacts from these effects should already be incorporated into the design due to the V_{ref} uncertainty (remember, when choosing the noise margin, many of the effects that are difficult or impossible to simulate are included). Subsequently, it is important that the noise margin be appropriately decreased when simulating phase 3 effects. For example, if the initial estimate was that I/O level power delivery effects will cause 40 mV out of 100 mV total noise, the noise margin should be decreased to 60 mV when calculating timings with a power delivery model in place. Often, if the designer is confident that the noise margin accounts adequately for these effects, phase 3 is not necessary except as a double check.

The phase 3 analysis builds on phase 2 and adds in the following effects:

1. I/O level power delivery (Section 6.2)
2. Nonideal current return paths (Section 6.1)
3. Simultaneous switching noise (Section 6.3)

Typically, each of the eight corner conditions (they must represent the absolute corners of the design space determined with phase 2) defined in Section 9.4.1 are simulated with power delivery, SSN and return path models. It should be noted, however, that a clean design should not exhibit any nonideal return paths because they are very difficult to model and can significantly affect the performance of the system.

If the rest of the design has been done correctly (i.e., the power delivery system has been designed properly and no nonideal return paths are designed into the system), this stage of the design will affect the design minimally. That is, the excess timings due to the original noise margin assumptions should be greater than or equal to the timing impact of the phase 3 analysis after the noise margin has been decreased adequately.

9.5. DESIGN GUIDELINES

The output of the sensitivity analysis and the layout study are used to develop the design guidelines. The guidelines should outline every parameter of the design that is required to meet the timing specifications. These guidelines are used to produce a high-volume design that will operate under all manufacturer and environmental variations. A subset of the guidelines should include the following:

1. Maximum and minimum tolerable buffer impedance and edge rates
2. Tolerable differences between rising and falling edge rates and impedances for a given chip
3. Maximum and minimum package impedance and lengths
4. Required power/ground-to-signal ratio and optimal pin-out patterns for packages and connectors
5. Maximum and minimum socket parasitics and possibly height
6. Maximum and minimum termination values
7. Maximum and minimum connector parasitics
8. Specific topology options including lengths
9. Maximum stub lengths (this may be the same as the package length, or it could be a daughtercard trace)
10. Impedance requirements for the PCB, package, and daughtercard
11. Minimum trace–trace spacing

12. Minimum serpentine lengths and spacing
13. Maximum variation in lengths between data signals and strobes
14. Plane references (As discussed in Chapter 6)
15. I/O level power delivery guidelines (see Chapter 6)

9.6. EXTRACTION

After the system has been designed and the package and PCB board routed, the final step is to examine the layout and extract the worst-case nets for the eight corners. These specific corners should be fully simulated using the phase 3 techniques to ensure that the design will work adequately. The significance list should help choose the worst-case nets. However, the following attributes should be examined to help identify the worst-case nets:

1. Signal changing layers
2. Significant serpentine
3. High percentage of narrow trace–trace spacing
4. Longest and shortest nets
5. Longest stub
6. Longest and shortest package lengths
7. Impedance discontinuities

9.7. GENERAL RULES OF THUMB TO FOLLOW WHEN DESIGNING A SYSTEM

As high-speed digital systems become more complex, it is becoming increasingly difficult to manage timings and signal integrity. The following list is in no particular order, but if followed, will hopefully make the design process a little simpler:

1. Early discussions between the silicon and system designers should be held to choose the best compromise between die ball-out and package pin-out. Often, the die ball-out is chosen without consideration of the system design. The package pin-out is very dependent on the die ball-out, and an inadequate package pin-out can cause severe difficulties in system design. Many designs have failed because of inadequate package pin-out.
2. When designing a system, make certain that the reference planes are kept continuous between packages, PCBs, and daughtercards. For example, if a package trace is referenced to a V_{DD} plane, be certain that the PCB trace that it is connected to is also referenced to a V_{DD} plane. Often, this is not possible because of package pin-out. This has been the cause of

several failures. It is very expensive to redesign a board and/or a package to fix a broken design.

3. Be certain that a strobe and its corresponding data are routed on the same layer. Since different layers can experience the full process variations, this can cause significant signal integrity problems.

4. Be certain that any scripts written to automate the sweeping function are working properly.

5. Make a significant effort to understand the design. Do not fall into the trap of designing by simulation. Calculate the response of some simplified topologies by hand using the techniques outlined in this book. Make certain that the calculations match the simulation. This will provide a huge amount of insight, will significantly help troubleshoot problems, and will expedite the design process.

6. *Look at your waveforms*! This may be the most important piece of advice. All too often, I have seen engineers performing huge numbers of parameter sweeps, looking at the three-dimensional plots and defining a solution space. This does not allow for an intimate understanding of the bus and often leads to an ill-defined solution space, wasting an incredible amount of time. Furthermore, when a part is delivered that does not meet the initial specifications, or a portion of the design changes, the engineer may not understand the impact to the design. It is important to develop and maintain basic engineering skills and not use algorithms in place of engineering. Double check your sweep results by spot checking the waveforms.

7. Keep tight communication with all disciplines, such as EMI and thermal.

8. Plan your design flow from simple to complicated. Initiate your design by performing hand calculations on a simplified topology. Increase the complexity of the simulations by adding one effect at a time so that the impact can be understood. Do not move on until the impact of each variable is understood.

9. Compensate for long package traces by skewing the motherboard routing lengths. When matching trace lengths to minimize skew, it is important that the electrical length from the driver pad (on the silicon) to the receiver pad be equal. If there is a large variation in package trace lengths, it is usually necessary to match the lengths by skewing the board traces. Subsequently, the total paths from silicon to silicon are equal lengths.

CHAPTER 10

Radiated Emissions Compliance and System Noise Minimization

As digital system frequencies increase it will become increasingly difficult to minimize electromagnetic radiation from the system. This creates a developing problem for the digital system designer in designing products that will pass the Federal Communication Commission's, or other regulating agency's, emission standards, commonly known as the *EMI* (electromagnetic interference) *specifications*. The importance of passing EMI specifications often takes precedence over many other design or functional requirements of a product because unlike other requirements, the need to pass EMI criteria is rooted in a government-driven specification rather than inspired by a market demand or engineering principles. In fact, it is, unfortunately, quite common, for system designers to be forced to compromise their own system timing requirements in order to pass the EMI test and be allowed to sell a product.

The topic of EMI is particularly difficult to handle in an engineering context because there are usually far too many variables to handle quantitatively and the EMI solution can usually not be finalized until prototypes have already been built and measured for compliance and failure mechanisms. At this late stage of the design cycle, it is often very difficult and expensive to make changes. This often leads to desperate measures, cost and schedule overruns, or solutions that compromise the intended function of the product.

Numerical or closed-form predictions of radiated emissions tend to be limited in utility, and expert systems are useful only for very limited purposes and are also inherently flawed by the inability to provide the program with enough input. Although many efforts are under way to develop more useful tools, the tools of the day do not do much better in predicting or suggesting solutions than an EMI specialist does "eyeballing" the system. It is encouraging for system designers, however, that the EMI engineer's specialized knowledge tends to be mainly composed of a few basic principles, some general design rules, an ability to apply trial and error, and a capacity to know when to abandon a theory or a rule of thumb. The design engineer's knowledge of these principles can prevent a late-blooming catastrophe.

One can begin to realize the difficulty in making accurate quantitative predictions by considering the fact that even the simplest antenna shapes are very difficult to solve and usually are solved only by relying on assumptions or empirical measurement. It is no surprise, then, that in systems with thousands of traces, complex shapes, many cables, and other variables, the subject sometimes tends to fall closer to superstition and mumbo jumbo than do most topics in engineering. It is very frustrating to rely on best-known methods that may not work, particularly when the reasoning behind them (or lack thereof) is not evident to the designer. The intent of this chapter is to equip the reader with the basic understanding and practical skill necessary to best navigate the task of minimizing radiation in all stages of a design cycle from conception to product.

10.1. FCC RADIATED EMISSION SPECIFICATIONS

Several regulatory agencies exist around the world which concern themselves with electromagnetic compatibility (EMC). In the United States, the Federal Communications Commission (FCC) sets forth EMC standards. Among these EMC standards are both conducted emission (such as noise conducted back through the power grid) and radiated emission limits. This chapter is concerned primarily with radiated emissions since that is typically the dominant emission problem in high-speed digital design.

The FCC radiated emission limits for digital devices is shown in Appendix F. There are two types of digital devices, class A and class B, each with its own emission requirements. Class B devices are those that are marketed for use in a residential environment, and class A devices are those that are marketed for use in a commercial, industrial, or business environment. Note in Appendix F that the class A and class B emission requirements are specified at different measurement distances. There is one more requirement, the *open-box emission requirement*, pertaining to computer motherboards that are marketed as a component rather than inside a complete system with a chassis. The open-box emission requirement is an emission requirement that must be passed with the chassis cover removed. The open-box requirements are relaxed from the closed-box requirements by 6 dB. This means that if a motherboard is to be marketed as a component rather then included in a complete system, under current regulation no more then 6 dB should be anticipated from chassis shielding.

10.2. PHYSICAL MECHANISMS OF RADIATION

One can show from classical electromagnetic theory that any accelerated charge radiates. Thus, in a digital system, which must accelerate charge in order to create electrical signals, radiation is inevitable. However, the rest is not that simple. It is well known that even very simple antenna shapes are difficult to analyze. Even the case of geometrically perfect shapes and perfect conductors, such as the case of a straight wire excited in the middle (half-wave dipole antenna), has been the subject of research for many years and is solved only by making assumptions about the current distribution. Thus in a real system with complex shapes, a myriad of impedance paths, many frequency components, and many unknown, unpredictable, or nonmeasurable variables, one may easily imagine that the task of predicting, controlling, and diagnosing radiating emissions would be intractable. To complicate matters, borrowing from methods that worked in the past could lead to more difficulties, because rule-of-thumb design practices are particularly treacherous if the user does not know in the appropriate situations to apply them. This is because several techniques contradict each other, and often to fix one mode of radiation, you cause another mode to increase.

Straightforward accurate simulation of the radiated fields generated from a complex system is generally not feasible, and even when done, is only sufficient to augment the judgment of a human engineer. This reveals the encouraging fact that application of a few simple principles can significantly reduce the complexity of the problem.

10.2.1. Differential-Mode Radiation

Differential-mode radiation results from current flowing in a loop. The model for differential-mode radiation is usually based in equations derived for a small magnetic dipole where the circumference, or the length, of the loop is much smaller than the wavelength (wavelength = $\lambda = c/F$, where c is the speed of light) of the signal (i.e., $2\pi a = \ell \ll \lambda$, such as shown in Figure 10.1). The application of this to a larger loop, encountered frequently in real-life systems, will be explained later. The following equations are from the solutions

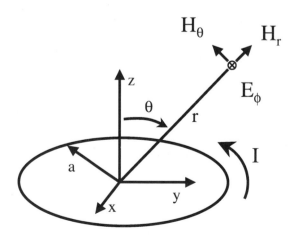

FIGURE 10.1   Small loop magnetic dipole.

of Maxwell's equations derived for an ideal magnetic dipole and are modified here to show only the field magnitude:

$$|E_\phi| = \frac{\eta_0 I A \pi}{r \lambda^2} \sqrt{1 + \frac{\lambda^2}{r^2 (2\pi)^2}} \sin \theta \qquad (10.1a)$$

$$|H_\theta| = \frac{I A \pi}{r \lambda^2} \sqrt{1 - \frac{\lambda^2}{r^2 (2\pi)^2} + \frac{\lambda^4}{r^4 (2\pi)^4}} \sin \theta \qquad (10.1b)$$

$$|H_r| = \frac{I A}{r^2 \lambda} \sqrt{1 + \frac{\lambda^2}{r^2 (2\pi)^2}} \cos \theta \qquad (10.1c)$$

where

E_ϕ = magnitude of the electric field, in volts per meter

H_θ = magnitude of the θ component of the magnetic field, in amperes per meter

H_r = magnitude of the r component of the magnetic field, in amperes per meter

$\eta_0 = 120\pi = 377\ \Omega$ = intrinsic impedance of free space $\left(\sqrt{\frac{\mu_0}{\varepsilon_0}} \right)$

λ = wavelength, in meters

I = current, in amperes

A = area of the loop, in square meters

r = distance from the center of the loop (see Figure 10.1)

θ = angle from z-axis (see Figure 10.1)

Similar solutions can be found in a multitude of electromagnetic textbooks for the interested reader. They are stated here to understand the basis of the practical approximations that will follow [Cheng, 1989; Mardiguian, 1992].

Notice that $\theta = \pi/2$ produces the worst-case field strength of E_ϕ and H_θ. Also notice that H_r falls off very rapidly with distance r from the center of the loop. The worst-case angle of $\theta = \pi/2$ will be assumed and H_r will be considered negligible for all that follows.

Differential Mode Near-Field Radiation. In conditions where the distance $r < \lambda/2\pi$, known as the *near-field*, terms with large exponents are dominant and the formulas reduce to

$$E_{\text{dm near}} = \frac{\eta_0 I A}{\lambda 2 r^2} = \frac{\mu_0 I A F}{2 r^2} = 6.3 \times 10^{-7} \frac{I A F}{r^2} \qquad \text{V/m} \qquad (10.2a)$$

$$H_{\text{dm near}} = \frac{I A}{4 \pi r^3} \qquad \text{A/m} \qquad (10.2b)$$

where

$E_{\text{dm near}}$ = differential mode near-field electric field strength, in volts per meter

$H_{\text{dm near}}$ = differential mode near-field magnetic field strength, in amperes per meter

$\eta_0 = 120\pi = 377 \ \Omega$ = intrinsic impedance of free space

μ_0 = permeability of free space = $4\pi \times 10^{-7}$ H/m

I = current, in amperes

A = area of the loop, in square meters

λ = wavelength, in meters

F = frequency, in hertz

r = distance from the center of the loop, in meters

Note that both the magnetic and electric components are strongly dependent on distance and the magnetic field is independent of frequency.

For future discussion, let us define the *wave impedance* to be the ratio E/H. The wave impedance for the near field of a loop radiator can be derived from (10.2a) and (10.2b) and is

$$Z_{\text{dm wave near}} = \frac{E}{H} = \frac{\eta_0 2 \pi r}{\lambda} \qquad \text{ohms} \qquad (10.3)$$

The wave impedance affects how strongly the wave will couple to other structures of given impedances, as discussed later in the chapter.

One may expect that the impedance of the wave in the vicinity of the circuit should be related to the circuit impedance, because, in fact, the circuit impedance itself is a result of the local fields (such as in $\sqrt{L/C}$, seen in other chapters as the impedance of a transmission line). With this in mind, examining equation (10.3) reveals that the impedance is low when λ is large. This approaches the dc impedance of the perfectly conducting uniform current loop in the model. Much like a perfectly conducting inductor, the impedance goes up in direct proportion to frequency (inverse proportion to wavelength).

Differential Mode Far-Field Radiation. If distances $r > \lambda/2\pi$, are considered, the terms with the large exponents will be negligible and the formulas will reduce to (10.4a) and (10.4b). The conditions where $r > \lambda/2\pi$ is known as the *far field* [Ott, 1988; Mardiguian, 1992; Cheng, 1989]

$$E_{\text{dm far}} = \frac{\eta_0 \pi I A}{\lambda^2 r} = \frac{\eta_0 \pi I A F^2}{c^2 r} = 1.32 \times 10^{-14} \left(\frac{I A F^2}{r} \right) \qquad \text{V/m}$$

$$(10.4a)$$

$$H_{\text{dm far}} = \frac{\pi I A}{\lambda^2 r} = \frac{\pi I A F^2}{c^2 r} \qquad \text{H/m} \qquad\qquad (10.4b)$$

Similar to before, the wave impedance can be calculated from the ratio of (10.4a) and (10.4b). The result is not surprising: $Z_{\text{dm wave far}} = \eta_0 = 120\pi$, which is the intrinsic impedance of free space.

Calculation for Nonideal Loops. Many references directly apply the equations derived above for applications that violate the initial assumption of $\ell \ll \lambda$ (loop length must be small compared to the wavelength). The reader should be aware that the ideal loop equations derived here are no longer theoretically correct when the condition of $\ell \ll \lambda$ is not met. At lengths larger than this, the circuit path will begin to behave like transmission lines and the current will no longer be uniform. For greater accuracy using the equations derived above assuming an ideal loop, the length of the loop should be less then approximately $\lambda/10$. However, for the purposes here, the equations can continue to be used for longer lengths if the following approximations are made. For total circuit loop lengths $\ell \geq \lambda/2$, the length ℓ should be clamped at $\ell = \lambda/2$ and the area A input into the equations above should be adjusted accordingly. For lengths $\ell < \lambda/2$, the actual loop area is used. For example, if the path is circular and the loop length is greater than $\lambda/2$, the corrected area A to be entered in the equations above should be calculated with a radius that produces a circumference of $\lambda/2$ [$C = \lambda/2 = 2\pi r \Rightarrow r = \lambda/4\pi$, thus $A = \pi(\lambda/4\pi)^2$]. If the path is a square rectangular loop, the corrected area A should be $(\lambda/8)^2 = \lambda^2/64$ (since the total loop length is $\lambda/8 + \lambda/8 + \lambda/8 + \lambda/8 = \lambda/2$). The physical reason for this "antenna shrinking" effect is that if the antenna is long enough compared to the frequency such that different peaks and valleys of the waveform exist on the antenna simultaneously, the various peaks and values will cancel each other's radiation [Mardiguian, 1992].

APPLICATION RULE

If the total loop path ℓ is longer than $\lambda/2$, the area A input into loop equations should be adjusted to the area that would be present with a maximum path of $\ell = \lambda/2$.

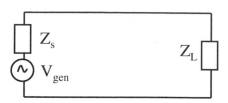

FIGURE 10.2   Nonideal loop that contains impedances.

As the reader may recall from equation (10.3), the near-field wave impedance was related to the circuit impedance assuming that the dc impedance was zero (perfectly conducting), which is not realistic. For the practical case, such as circuit loop that contains a source and load impedance (such as shown in Figure 10.2), the ideal assumption breaks down. The following equation calculates the near field for the nonideal case (when the total circuit impedance of the loop $Z_c = Z_s + Z_L \geq 7.9 \text{ rF} \times 10^{-6}$).

$$E_{\text{dm near } Z_c} = \frac{VA}{4\pi r^3} \qquad \text{V/m} \qquad \text{when} \quad Z_c \geq 7.9 \text{ rF} \times 10^{-6} \qquad (10.5)$$

Equation (10.2a) is valid if the impedance is less then $7.9 \text{ rF} \times 10^{-6} \ \Omega$. Furthermore, the modification is not required for the far field regardless of total loop circuit impedance. Note that the equation is written in terms of voltage instead of current. The other differential-mode equations may be written for voltage by approximating $I = V/Z_c$ [Mardiguian, 1992]. Some practical applications of (10.2), (10.3), (10.4), and (10.5) are explored later in the chapter.

Problematic Frequencies. Let us make a general approximation of which frequencies are of most concern in a digital system for differential-mode radiation. Consider Figure 10.3, which shows a Fourier spectrum of a trapezoidal clock signal and the associated envelope that typically bounds the spectrum of a digital signal. For a perfectly symmetric (i.e., 50% duty cycle) digital signal, one can verify with Fourier analysis, no even harmonics will be present in the system. However, in reality there are sometimes significant even harmonics.

Note that the frequency envelope of the Fourier spectrum shown in Figure 10.3 falls off with frequency at a rate of -20 dB per decade (an inverse linear relationship with frequency) up to $f = 1/\pi T_r$ after which it falls off at a rate of -40 dB per decade (an inverse-square relationship).

Now consider the far-field radiation [equation (10.4a)], which is directly proportional to the square of frequency and thus increases with frequency at a rate of $+40$ dB per decade. However, when the loop path is greater then $\lambda/2$, the effective loop area is adjusted to be a constant times λ^2 (the constant will depend on the area calculation of the shape of the loop). Thus at high enough frequency, the λ^2 component of the area A in equation (10.4a) will cancel with the λ^2 element, and the field will stay constant with increasing frequency. This will happen at a frequency $F = c/2\ell$ (where c is the speed of light and

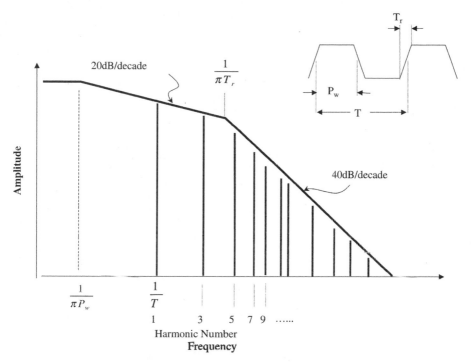

FIGURE 10.3   Spectrum of a digital signal showing peak envelope regions.

ℓ is the loop length). Figure 10.4 shows the relative radiated magnitudes of different frequencies input into a given loop. Note that the loop will radiate frequencies with increasing efficiency up until the antenna shrinking effect occurs at frequencies above $F = c/2\ell$.

When the spectral envelope of the digital signal (the input into the radiator) is combined with the frequency dependence of the far-field radiation shown in Figure 10.4, we arrive at the differential radiated emission envelope shown in Figure 10.5. Note that the frequency $F = 1/\pi T_r$ will not necessarily occur at a higher frequency than $F = c/2\ell$, as shown in the figure. The frequencies that will be most likely to radiate (differential mode) are contained between the frequencies $F = 1/\pi P_W$ and $F = 1/\pi T_r$ or $F = c/2\ell$ (whichever is highest).

RULE OF THUMB: Minimizing Differential-Mode Radiation

Since the magnitudes of the fields are proportional to the loop area, smaller loops produce less radiated emission. During design, minimize current loops whenever possible.

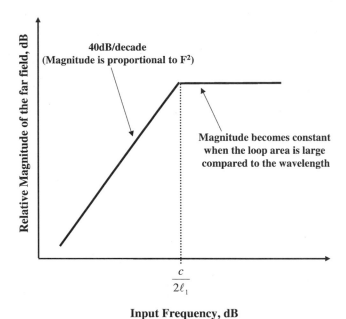

FIGURE 10.4   Relative magnitude of far-field radiation as a function of frequency.

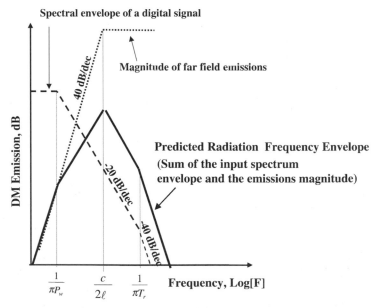

FIGURE 10.5   Resultant differential radiation behavior when far-field characteristics are combined with spectral input.

10.2.2. Common-Mode Radiation

Common-mode currents are those flowing in the same direction. A common source of common-mode currents in a digital system is from noise on ground or power planes due to the switching noise of digital circuits. Common-mode currents are likely to cause emission concerns when they find access to external cables, which can act as long antennas. Often, the common-mode effects determine the overall emission performance.

One may attempt to avoid common-mode current altogether by providing return paths for all signals (such as the shield on a cable or by providing a very good ground return path for a signal), in which case one might expect common-mode current to cancel. However, there are always unanticipated return paths that cause some of the return current to deviate from the intended path. This will cause enough asymmetry to cause a net common-mode effect. Figure 10.6 illustrates a simple example of the generation of common-mode current. It takes only a very small amount of current deviating from the expected complete loop to cause significant common-mode emissions.

The following solutions of Maxwell's equations for an ideal electric dipole are stated here to provide an understanding of common-mode radiation:

$$|E_r| = \frac{60I\ell}{r^2}\sqrt{1 + \frac{\lambda^2}{r^2(2\pi)^2}}\cos\theta \qquad \text{V/m} \tag{10.6a}$$

$$|E_\theta| = \frac{I\ell\eta_0}{2\lambda r}\sqrt{\frac{\lambda^4}{r^4(2\pi)^4} - \frac{\lambda^2}{r^2(2\pi)^2} + 1}\sin\theta \qquad \text{V/m} \tag{10.6b}$$

$$|H_\phi| = \frac{I\ell}{2\lambda r}\sqrt{1 + \frac{\lambda^2}{r^2(2\pi)^2}}\sin\theta \qquad \text{A/m} \tag{10.6c}$$

where ℓ is the length of the radiating wire. An electric dipole is illustrated in Figure 10.7. An electric dipole is essentially equivalent to an isolated wire in space with the condition that $\ell \ll \lambda$. The bulbs at the ends of the line shown in Figure 10.7 are included so that a uniform current can be assumed on the small length of the conductor, due to the capacitance of the bulbs. The interested reader can find similar solutions to an elemental electric dipole in many electromagnetic textbooks [Cheng, 1989]. As before, these equations have been modified to show only field magnitude.

Notice that (10.6b) and (10.6c) are maximum for $\theta = \pi/2$. As before, this condition will be assumed from now on for worst-case field emissions. E_r falls off more rapidly with distance (inverse r^2) from the wire and is neglected in the following approximations.

Common-Mode Near-Field Radiation. For distances r less then $\lambda/2\pi$, the terms with the higher exponents will be dominant, so equations (10.6b)

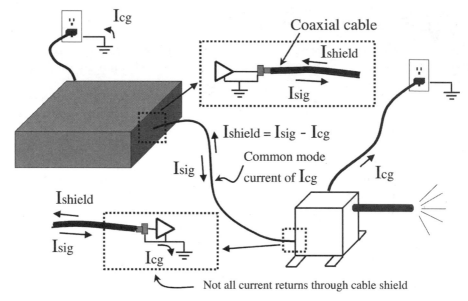

FIGURE 10.6 Common-mode current generation.

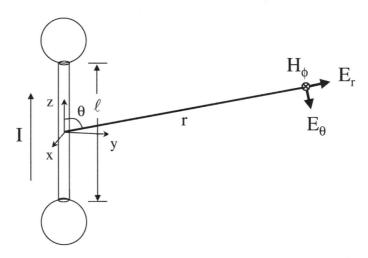

FIGURE 10.7 Illustration of an electric dipole.

and (10.6c) reduce to

$$E_{\text{cm near}} = \frac{I\ell\eta_0\lambda}{8\pi^2 r^3} = 4.8\left(\frac{I\ell\lambda}{r^3}\right) \qquad \text{V/m} \qquad (10.7a)$$

$$H_{\text{cm near}} = \frac{I\ell}{4\pi r^2} \qquad \text{A/m} \qquad (10.7b)$$

As before, this region is known as the *near field*. Note that the peak H field is independent of frequency. This equation, in fact, holds down to dc. As before, we can produce the common-mode near-field wave impedance as the ratio of (10.7a) and (10.7b). E/H is shown in the equation

$$Z_{\text{cm wave near}} = \frac{\eta_0 \lambda}{2\pi r} \quad \text{ohms} \tag{10.8}$$

As before, the wave impedance close to the radiator can be expected to relate to the circuit impedance. Note that for $\lambda = \infty$, the impedance is infinite, as one would expect for an open-circuited wire.

Common-Mode Far-Field Radiation. When distances r greater then $\lambda/2\pi$ are considered, equations (10.6b) and (10.6c) reduce to

$$E_{\text{cm far}} = \frac{\eta_0 I \ell}{2\lambda r} = 6.28 \times 10^{-7} \left(\frac{\ell I F}{r} \right) \quad \text{V/m} \tag{10.9a}$$

$$H_{\text{cm far}} = \frac{I \ell}{2\lambda r} \quad \text{A/m} \tag{10.9b}$$

This region is known as the *far field* [Ott, 1988; Mardiguian, 1992; Cheng, 1989]. Not surprisingly, the wave impedance in this region can be seen [from the ratio of equations (10.9a) and (10.9b)] to be the impedance of free space, η_0.

Nonideal Dipole Radiators. As for the radiating loop, many references directly apply electric dipole equations for frequencies and circuit dimensions that violate the initial assumption of $\ell \ll \lambda$. The reader should be aware that the solutions are not theoretically correct when this assumption is not met. However, equations (10.7a) and (10.9a) can be modified to be usable with longer dimensions, much like the length of the loop was modified for the loop radiator, by fixing the length of the radiating wire to $\lambda/2$. Beyond this length, the radiating wire looks like a transmission line and the current is not uniform. Thus the effective radiating area shrinks. The fields from the other $\lambda/2$ segments cancel each other out [Mardiquian, 1992].

If only one side of the radiating wire has a strong capacitance or connection to ground, the radiating wire will behave like a monopole rather then a dipole, and twice the length should be used in equation (10.9a). It should be noted that at resonant lengths (such as multiples of $\lambda/2$) where standing waves will be present on the radiator, there will be directional lobes to the radiation, and the maximum field intensity may be significantly higher.

Common-Mode Radiation in the Presence of a Reference Plane. Another modification is required for a radiating wire if the radiating wire is close to a reference plane (ac ground). A ground plane that extends sufficiently in the vicinity of the radiator (i.e., a quasi-infinite plane) will act as a reflector. The reflected wave will be inverted from the incident wave much like a wave on a transmission line reflecting from a short-circuit termination. Depending on the distance of the reflector and the wavelength being considered, the reflected wave can add constructively or destructively with the radiated wave.

To see the destructive and constructive effects of the reflecting plane, one may use trigonometric identities as follows. Consider the radiated wave to be $\sin\theta$ and the phase shift angle of the reflected wave to be $\sigma = 2\pi\lambda/h$, where h is the distance from the constant-voltage plane. Thus the resultant wave will be $\sin\theta - \sin(\theta + \sigma)$. Substituting the trigonometric identity $\sin(\theta + \sigma) = \sin\theta\cos\sigma + \cos\theta\sin\sigma$ reveals that if the phase shift is small, such as in the case of a PCB trace over a ground plane, then $\sin\sigma \approx 0$, $\cos\sigma \approx 1$, and the incident wave will cancel with the reflected wave, causing very little radiation. This is why PCB board traces are typically not considered for common-mode radiation except for very high frequencies or traces with no local reference plane. In such a case, the loop radiator equation can still be applied to the board trace. In a case where the phase shift of the reflected wave is larger, say close to a value of π, then $\sin\sigma \approx 0$, $\cos\sigma = -1$, and the reflected and radiated waves add constructively to *double* the radiated amplitude.

For a *close* ground plane (closer then $\lambda/10$), the reduction in radiation is sometimes given as [Mardiguian, 1992]

$$\frac{E_{\text{with gnd}}}{E_{\text{no gnd}}} = \frac{10h}{\lambda} \tag{10.10}$$

where h is the distance from the reference plane. Factoring equation (10.10) with equation (10.9a) yields

$$E_{\text{with gnd}} = \frac{\eta_0 I\ell}{2\lambda r}\frac{10h}{\lambda} = \frac{10h\eta_0 I\ell}{2\lambda^2 r} = \frac{10\eta_0 I A F^2}{2c^2 r}$$

$$= 2.1 \times 10^{-14}\left(\frac{IAF^2}{r}\right) \quad \text{V/m} \quad \text{for} \quad \frac{10h}{\lambda} < 1 \tag{10.11}$$

Notice that this is very similar to the loop radiation equation (10.4a).

Problematic Frequencies. As before, let us determine the emission envelope of most concern. Equation (10.9a) is directly proportional to F and thus will increase with F at a rate of 20 dB/decade. Recalling the digital spectrum envelope as shown in Figure 10.3, and recalling the long wire modification, the envelope shown in Figure 10.8 can be constructed in a manner similar to Figure 10.5. Subsequently, the frequencies that will be most likely to radiate significantly are contained between the frequencies $F = 1/\pi P_W$ and $F = 1/\pi T_r$ or $F = c/2\ell$ (whichever is highest).

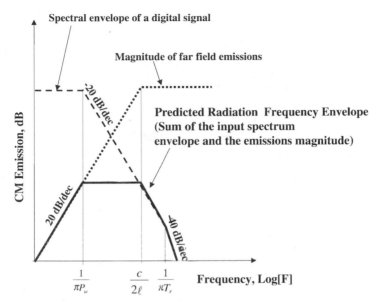

FIGURE 10.8 Resultant common-mode radiation behavior when far-field characteristics are combined with spectral input.

10.2.3. Wave Impedance

The wave impedance, the ratio E/H, was mentioned above for each type of radiation. The wave impedance can be thought of as an impedance just like the impedance of a transmission line. Thus, whether the radiated wave will couple to other mechanisms can be understood based on the impedance of the structures. For instance, a low-impedance wave, such as the near-field portion of a loop radiator, will couple to a conductive (low-impedance) shield. A high-impedance wave, such as the near field of a common-mode radiator will be reflected from a conductive shield due to the impedance mismatch between the high-impedance wave and the low-impedance shield. For all types of radiation, the far-field wave impedance eventually approaches the impedance of free space, $\eta_O = 377\ \Omega$. The wave impedance is shown graphically in Figure 10.9.

Because emission requirements are typically specified in terms of electric field, there are also dependencies on the measured distance (since a low-impedance radiator in the near field will look magnetic and not show up on the electric field probes). A common practice is to take measurements at a given distance and back-calculate to the distance specified for the emission requirement. This could lead to inaccuracies, depending on the wave impedance.

Understanding the nature of wave impedance can also be a useful diagnostic. For example, if the magnitude of the electric field appears to get larger with increasing distance, then the radiator is probably a loop. To understand this, refer to Figure 10.9. Notice that in the far field, the impedance for the

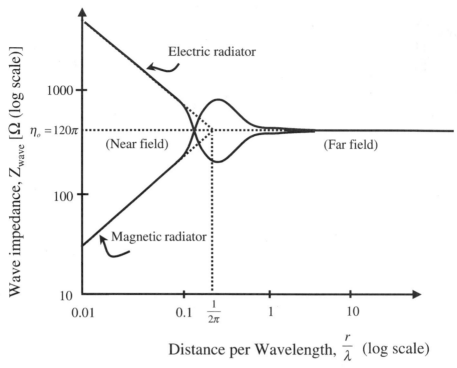

FIGURE 10.9   Wave impedance behavior.

electric and the magnetic radiator both assume the value of 377 Ω. This means that the electric field probes used to measure the far-field radiation cannot distinguish between a magnetic and an electric radiator. However, as the probe is moved towards the radiator and into the near field, the measured electric field will decrease with distance for a loop radiator because the radiation becomes primarily magnetic as the wave impedance decreases. Obviously, the opposite also holds. If the magnitude of the electric field increases as the probe is moved towards the radiator, then the radiator is probably due to common-mode currents. Alternatively, if the measured radiation from a problematic frequency in the far field is relatively unaffected by placing a low impedance metal shield in the near field (such as a computer chassis), then this is a clue that the radiator is a loop. If the shielding significantly reduces the magnitude of the radiation, the radiator is probably due to common-mode currents.

10.3. DECOUPLING AND CHOKING

This section may well extend far beyond the motives of this chapter. The simple task of decoupling a high-frequency digital circuit is paramount to signal

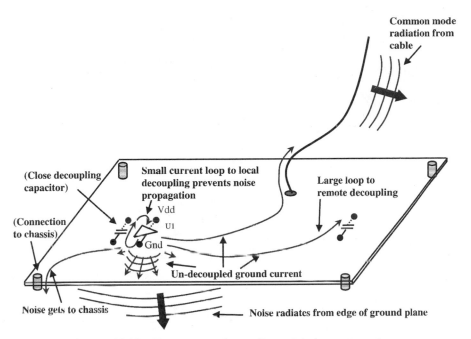

FIGURE 10.10 How proper decoupling minimizes current loops.

integrity, power delivery, and minimization of emissions. Decoupling amounts to creating low ac impedance between power and ground structures (although in some circumstances, the exact opposite is done to minimize emissions, this having a negative effect on signal integrity and timings). The importance of decoupling in reducing emissions can be seen from Figure 10.10. In the figure it can be seen that the presence of the decoupling capacitor prevents the current from taking a much larger loop path or from finding its way to a cable and radiating as a common-mode radiator. Figure 10.10 suggests not only that the local decoupling capacitor value should be chosen appropriately but that the capacitor is placed physically close to the component to minimize the loop area. If current cannot be decoupled appropriately, inductive choking may be used to block the passage of high-frequency noise. Inductive components are regularly used at cable connections to minimize common-mode currents. Occasionally, regions of the reference planes are isolated and choked off with an inductor, as shown in Figure 10.11. Cutting the reference planes should generally be avoided in high-frequency digital systems unless other methods have been exhausted, for reasons explained in this section. Understanding decoupling or choking is not sufficient without an understanding of a few limitations, particularly at high frequency. Several modifications of the examples presented here exist and are almost universally applicable.

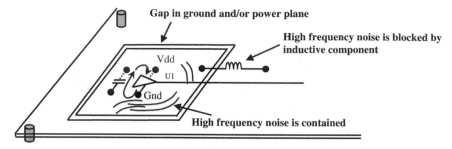

FIGURE 10.11   How inductive choking minimizes current loops.

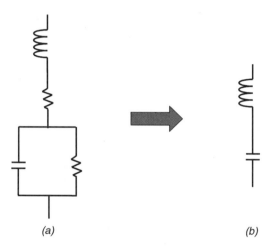

FIGURE 10.12   Decoupling capacitor: (*a*) real model; (*b*) model used here.

10.3.1. High-Frequency Decoupling at the System Level

The model for a real-life capacitor is shown in Figure 10.12. Some impedance curves of typical capacitors in typical packages are shown in Figure 10.13. Note that each capacitor has a resonance at which its impedance is at a minimum. The effective resonance that forms the minimum impedance point when placed on a circuit board will be slightly different due to the added inductance in the path routed to the capacitor. Note also that the choice of capacitor dielectric can lower the impedance of a capacitor.

The idea behind decoupling is, of course, to provide a minimum ac impedance between power and ground. Thus one may conclude by a glance at Figure 10.13 that a few higher-value capacitors chosen for low impedance at low frequency paired with some lower-valued capacitors chosen to provide low impedance at high frequencies (near their resonance points) would be the method of choice. This is more or less the optimal way to decouple a

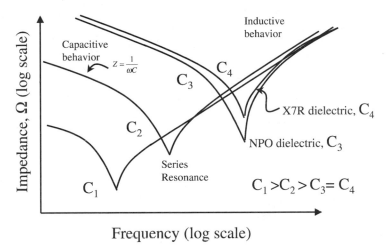

FIGURE 10.13 Frequency dependence of a decoupling capacitor.

board; however, there are other effects that are not immediately obvious. For example, resonance conditions can exist such that the decoupling capacitors can actually worsen the impedance of the power delivery for some frequencies. This can happen when the total inductance associated with the capacitors combines in parallel and resonates with the plane capacitance. This can actually cause the impedance between power and ground to increase to beyond what the impedance would be with no decoupling capacitors present. To see this, let us a build up a model of a decoupled power and ground plane.

Because the inductance of a power or ground plane is very small, let us assume it to be zero. This allows us to model the power and ground plane structure as a capacitor with no series inductance. Typically, the plane capacitance is on the order of 100 pF/in^2 for *close* power and ground planes (10 mils). These close planes are typically only implemented on boards that have more than four layers, and probably at least eight layers. We will see shortly some benefits of having close planes. It should be mentioned that spacing the planes very close together (i.e., closer then 4 mils) may be a reliability issue with the board manufacturer because of the difficulty in handling very thin sheets of dielectric. The distance between planes on a much more typical four-layer board will probably be more on the order of 45 mils. The stackup can be examined to determine this spacing. The plane capacitance may be estimated, if necessary, by the familiar parallel-plate capacitor equation $C = \varepsilon_0 \varepsilon_r A/d$ using $\varepsilon_r = 4.2$ (nominal) for the dielectric constant of FR4. Considering that even the high frequency of 1 GHz has a wavelength of approximately 6 in. in an FR4 medium ($\lambda = c/F\sqrt{\varepsilon_r}$), it is fairly reasonable for system-level decoupling purposes to model the ground plane as a lumped capacitor.

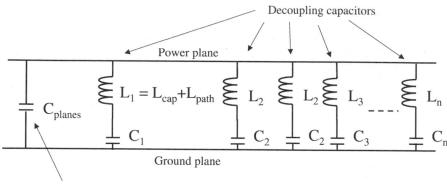

FIGURE 10.14 Simplified model of a ground plane with decoupling capacitors.

A refined model would include a distributed capacitance, inductance, and perhaps losses. If the simplified model is accepted and the *LC* elements of the decoupling capacitors are entered, the resulting model is as shown in Figure 10.14. Each inductance in the figure will be different, due to the differing paths to the decoupling capacitors. However, for now let's assume that the path inductances are equal, so that $L_1 = L_2 = \cdots = L_n$ and $C_1 = C_2 = \cdots = C_n$. Under this condition, the following equation for the impedance of the board can be derived [Hubing et al., 1995]:

$$Z_{\text{board}} = \frac{j(\omega L_1/n - 1/n\omega C_1)}{1 - \omega C_{\text{planes}}(\omega L_1/n - 1/n\omega C_1)} \tag{10.12}$$

From equation (10.12) it can readily be derived that the impedance will be zero for the condition of

$$F_{\text{series}} = \frac{1}{2\pi\sqrt{L_1 C_1}} \tag{10.13}$$

At this point the capacitors are resonating with their own self-inductance plus the path inductance. This point is similar to the minimum impedance point seen in the capacitor curves in Figure 10.13 except that there is an added inductive component in L_1 from sources other then the capacitor itself. (The inductance of the capacitor itself is typically 1 nH for common surface-mount capacitors.) At frequencies higher than the series resonant frequency, the slope of the impedance curve for the given capacitor changes, meaning that the circuit has become inductive. Thus the circuit will present more impedance with higher frequency. Thus it is evident that the series inductance should be kept to a minimum. To put this in perspective, even a very close and low inductive path to a nearby decoupling capacitor on a standard PCB board can easily be 2 nH. One standard via is approximately 0.7 nH; thus, if it takes two vias to connect a decoupling capacitor (one via down to the power plane, and one on the ground side of the capacitor), the inductance is 1.4 nH. This is assuming no inductance from the plane or any other source, such as a package power or ground pin. Thus, even a well-placed capacitor on a standard board may have a total series inductance of 3 nH (1.6 nH for the capacitor self-inductance). Assuming that the capacitance of the decoupling capacitor should be significantly more than the capacitance of the planes, with a 6 in. by 6 in. board with close power and ground planes (10 mils), this means that the value of the decoupling capacitance should be larger than approximately 4 nF. Calculating the resonance using a 6-nF capacitor with 3 nH of inductance reveals a zero at 38 MHz, above which the capacitor will become less and less effective. This is discouraging for high-frequency decoupling. The mistake should not be made, however, that the capacitor is necessarily completely ineffective above the series resonance simply because the inductance is dominating the impedance. As long as the total impedance is less than the impedance without the capacitor present, the capacitor is providing some decoupling.

It can also be seen from equation (10.12) that the impedance will be *infinite* at the condition met in the equation

$$F_{\text{parallel}} = F_{\text{series}} \sqrt{1 + \frac{nC_1}{C_{\text{planes}}}} \tag{10.14}$$

This is a pole in the impedance equation. At this point, the *plane* capacitance is resonating in parallel with the combined series inductance of the decoupling capacitors. Note that increasing the number of decoupling capacitors increases the frequency of this pole. At this frequency, current cannot be drawn from the power/ground structure. Thus, decoupling has not only become ineffective, but has actually become destructive to the goal. This can be seen in Figure 10.15, which shows a plot of the impedance of the bare board (modeled as a pure capacitor) superimposed on the impedance of the board loaded with decoupling capacitors.

The frequency at which the impedance of a bare board is the same as the board with the decoupling capacitors is given by the equation [Hubing et al., 1995]

$$F_A = F_{\text{series}} \sqrt{1 + \frac{nC_1}{2C_{\text{planes}}}} \tag{10.15}$$

Beyond this frequency, the bare board impedance is *less* then the board loaded with decoupling capacitors.

The equations above can be derived assuming that all inductance and capacitance values of the decoupling capacitors are equal. In reality, of course,

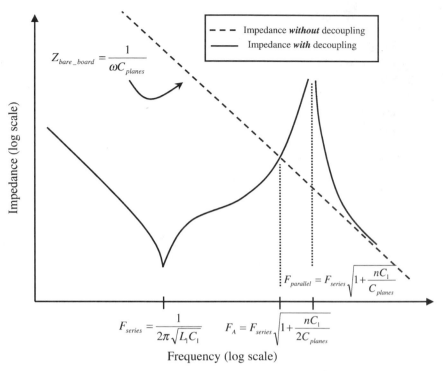

FIGURE 10.15   Poles and zeros of a board populated with decoupling capacitors.

the values differ. There will, in fact, usually be several poles and zeros, due to the different series and parallel resonances for different inductance and capacitance values present. Equation (10.14) does not work for differing decoupling capacitors and/or differing inductance values. However, to estimate the general shape of the impedance curve, it is helpful to note that there is one parallel resonance between each pair of series resonances.

Two example cases are plotted in Figure 10.16 with log-log scales. Superimposed on the plot is the impedance curve of the bare board if it were not populated with any decoupling capacitors. Very important, note that simply increasing the number of capacitors increases the frequency of the point F_A (beyond which decoupling is ineffective). Thus a large *number* of capacitors is important to have in the system, preferably located close to the component to be decoupled. There will be distant capacitors (such as capacitors surrounding different components) that will contribute to the total decoupling; however, the inductance of the power plane may not be negligible at far distances. Furthermore, in the interest of minimizing loop area it is preferable to have the decoupling local.

Due to the series inductance inherent in the use of discrete capacitors, the decoupling network for a motherboard will become ineffective for frequen-

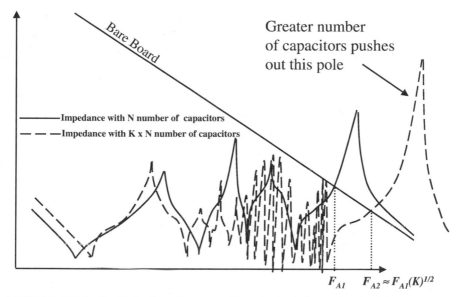

FIGURE 10.16 Realistic impedance profile of a board populated with decoupling capacitors.

cies that approach the pole described by equation (10.14). Depending on the implementation, this frequency can occur as low as 300 or 400 MHz.

A current area of study involves embedded plane capacitance in the board itself. Not surprisingly, these studies have shown far superior decoupling for these systems because the series inductance is very small. There are also several low-inductance capacitors available in the industry, such as "reverse" form factor surface-mount or interleaved capacitor packs. These devices are already used in some high-volume applications. However, use of these capacitors require creatively minimizing the inductance of the placement on the circuit board, or the benefit is lost.

10.3.2. Choking Cables and Localized Power and Ground Planes

In Section 10.3.1 it was noted that decoupling high-frequency components might be infeasible. This situation creates an environment in which high-frequency noise propagates unhindered through a system creating a high probability that the noise will find an efficient radiator. Consider a high-frequency driver that induces noise onto the power and ground planes such as shown in Figure 10.10. Since the high-frequency noise is not decoupled, it is free to propagate through a large number of paths. Often, the noise will access an external cable and radiate as a common-mode radiator. For this reason, inductive components are sometimes used to choke off the high-frequency

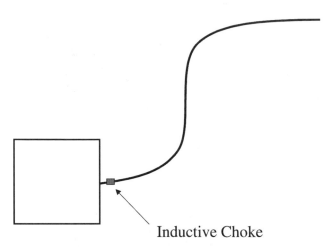

Inductive Choke

FIGURE 10.17   Choking off noise coupled onto a cable.

noise. This is commonly done at cable connection points, as illustrated in Figure 10.17 using board-mounted devices or ferrites that surround the outside of the cable. This technique is also sometimes used to choke off local power and/or ground regions containing noise sources as shown in Figure 10.11. The inductive components used for such purposes are usually ferrites, which are lossy inductive components. The losses help minimize the magnitude of possible resonances as well as dissipate some of the noise energy.

When reference plane isolation is applied, the power and/or ground connections to the planes are made with an inductive component. The inductance will choke off high-frequency components. As described in Chapter 6, however, it is optimal from a signal integrity point of view to minimize the inductance to the power supply. Subsequently, there can be significant trade-offs in signal integrity and high-frequency power delivery capability. By solving an immediate problem, this method often creates different problems in signal integrity, power delivery, and even increases some other types of emissions. First, the inductive power can make it impossible for choked off components to operate in specification by starving the component from high-frequency power. Subsequently, the power/ground environment seen by the component will be much more noisy (see Chapter 6). Second, creating islands in the power or ground planes may imply that signal lines must cross over a gap in the reference plane, which causes other significant signal integrity problems (see Chapter 6).

The method of localizing and choking off reference planes should not be excluded from consideration, but its trade-offs should be understood. Choking mechanisms for both external cable connections and, when absolutely necessary, for isolated power/ground regions are explored in the following two sections.

Choking Cable Entry Points. A wide variety of inductive components are available to minimize common-mode noise on cables. These products include ferrites that fit around the cable itself (including ferrites that are deliberately split, which are meant to be retrofitted easily to a cable that is already in place), board-mounted ferrites, and even cables with ferrite material integrated. These components are available in high- or low-loss versions but are typically lossy. These ferrites work based on their physical property of high permeability μ_r (one manufacturer lists a range of 15 to 1800). The most typical application involves physically slipping a ferrite around the outside of a cable, thus effectively increasing the effective permeability seen by the cable and thus also increasing the series inductance. However, the inductance is seen only for common-mode currents because the magnetic fields from opposite currents will cancel (unless the two currents pass through different holes in the ferrite, as in multihole ferrites). For differential signals or a coaxial cable, the desired signals are not affected as long as the anticipated return path is being utilized. Thus, except for applications in which the current is flowing deliberately in only one direction, the ferrites affect only the undesired components of the current.

Although ferrite beads are generally thought of as inductors, they are in fact lossy transformers, which is consistent with their intended purpose of dampening high-frequency components. The impedance that the ferrite introduces can be modeled as a resistive component and an inductive component, as shown in the equation [Mardiguian, 1992]

$$Z_{\text{cm ferrite}} = \sqrt{R^2 + \mu_e^2 L_0^2 \ell^2 \omega^2} \qquad (10.16)$$

where L_0 is the inductance per length of the cable or wire under consideration, μ_e the effective relative permeability of the ferrite (listed by the manufacturer, typically a few hundred), ℓ the length of the cable covered by the ferrite, and R the resistive (loss) component of the ferrite. The reader may have deduced from (10.16) that the inductance offered by the ferrite is

$$L = \mu_e L_0 \ell \qquad (10.17)$$

For a coaxial cable, the inductance per length, L_0, can be estimated from $(\mu/2\pi)\ln(r_{\text{shield}}/r_{\text{inner conductor}})$. Since the cable usually contains no magnetic material, the permeability is usually equal to that of free space, $\mu_0 = 4\pi \times 10^{-7}$ H/m. If the ratio of the radii of the coaxial cable is not known, they can be solved from the impedance formula $Z_0 = (\sqrt{\mu/\varepsilon}/2\pi)\ln(r_{\text{shield}}/r_{\text{inner conductor}})$, which, except for cases with $\mu_r \neq 1$, usually results in $Z_0 = 60(\varepsilon_r)^{-1/2}\ln(r_{\text{shield}}/r_{\text{inner conductor}})$. If the dielectric of the cable is unknown, usually Teflon can be assumed with $\varepsilon_r \approx 2.1$. For 50-$\Omega$ cable, the ratio comes out to be about 3.6 [Ramo et al., 1994].

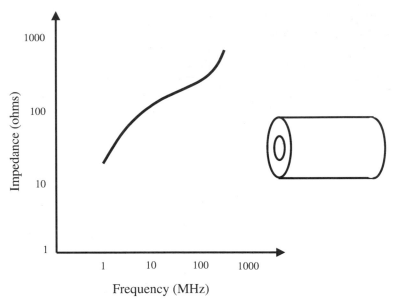

FIGURE 10.18 Impedance variation of a ferrite core. Ferrite dimensions: inner diameter, 8.0 mm; outer diameter, 16 mm; length, 28.5 mm.

The impedance of the ferrite core is typically listed by the manufacturer and ranges from about 50 to 600 Ω, more commonly in the range 100 to 200 Ω. The impedances are generally listed at 25 and 100 MHz; however, the impedance curve is generally more complex and increases steeply with frequency. A typical impedance curve of a cylindrical ferrite used to slip over the outside of a cable is shown in Figure 10.18. Such curves, which are provided by the manufacturer, should be used to determine the ferrite impedance at the desired frequency. The attenuation of the ferrite is

$$\alpha_{\text{ferrite}} = 20\log \frac{V_{\text{with ferrite}}}{V_{\text{without ferrite}}}$$

$$= 20\log \frac{Z_L/(Z_{\text{cable}} + Z_s + Z_L + Z_f)}{Z_L/(Z_{\text{cable}} + Z_s + Z_L)}$$

$$= 20\log \frac{Z_{\text{cable}} + Z_s + Z_L}{Z_f + Z_L + Z_s + Z_{\text{cable}}} \tag{10.18a}$$

where, Z_s, Z_L, and Z_{cable} are the source, load, and ac resistance of the cable, respectively, and Z_f is the impedance of the ferrite. For example, if a choke such as that depicted in Figure 10.18 is slipped around a cable and the source and load impedance values are 50 Ω and the losses of the cable are neglected, the resultant best case attenuation at 100 MHz induced from the ferrite is approximated as

$$\alpha(100 \text{ MHz})_{\text{ferrite}} = 20\log \frac{Z_{\text{cable}} + Z_s + Z_L}{Z_f + Z_L + Z_s + Z_{\text{cable}}}$$

$$= 20\log \frac{0 + 50 + 50}{250 + 50 + 50 + 0} = -10.8 \text{ dB} \quad (10.18b)$$

The ferrite impedance at 100 MHz is determined to be approximately 250 Ω from the chart in Figure 10.18. It can be seen that the ferrite can have a significant effect on common-mode currents and thus be very useful in prevention of common-mode radiation. Ferrites can be used with internal as well as external cables. It should be noted that if the ferrite is to be used for external cables on devices with a metal case, they should be placed as close as feasible to the actual boundary of the case of the device. If the ferrite is placed a distance inside the metal case, the cable can pick up noise after it passes through the ferrite. If the ferrite is placed a distance from the outside of the metal case, there will be a segment that can act as a good common-mode radiator.

One can improve the function of the ferrite by making multiple turns of the cable through the ferrite. However, this may bring the ferrite into saturation and reduce the inductive effects of the ferrite. Furthermore, the turn-to-turn capacitance may provide a bypass for high-frequency components and negate the effect of the ferrite. The increase in impedance of the ferrite is proportional to the N^2, where N is the number of turns. A condition of saturation can be checked by using the equation

$$B = \frac{\Phi}{A_{\text{flux}}} = \frac{LI}{A_{\text{flux}}} = \frac{\mu_e L_0 \ell I}{\ell(r_{\text{outer}} - r_{\text{inner}})} = \frac{\mu_e L_0 I}{r_{\text{outer}} - r_{\text{inner}}} \quad \text{tesla} \quad (10.19)$$

to check the flux density, B. One manufacturer lists saturation levels ranging from 2800 to 3800 G (units of gauss must usually be converted to units of tesla; 1 T = 10^4 G). Using a typical permeability of a ferrite, $\mu_e = 1500$, a maximum flux density of 3000 G (as provided in the data sheet), an inductance per length of a RG58/U cable (found as described above) of 260 nH/m, and the ferrite pictured in Figure 10.18, the maximum current before the ferrite will saturate is found as follows:

$$B_{\text{max}} = 3000 \times 10^{-4} = 1500 \times 260 \times 10^{-9} I_{\text{max}} \frac{1}{(0.016 - 0.008)\frac{1}{2}}$$

$$I_{\text{max}} = 3.1 \text{ A}$$

Certainly for most applications this level of current will not be approached. The exceptions may be in power cables or in large parallel data cables. If multiple turns are made through the ferrite, the maximum current must be derated by N^2.

Inductive Isolation, Cutting Ground/Power Planes, and Star Grounding.
A method sometimes employed to minimize the effect of a noisy component coupling noise into a system where it may be radiated is to physically cut the ground and/or power reference planes and feed the localized power and/or ground through an inductive component. The idea behind this method is to isolate the high-frequency noise components generated by the device and prevent them from propagating through the system. In digital systems historically, this has frequently been done with the system clock chip. This is illustrated in Figure 10.11. The method can be applied to just one plane or both the power and ground plane for a noisy component.

As mentioned previously, using this method has several detrimental effects in a high-frequency digital system. Generally speaking, this method should not be considered unless other methods are exhausted or prior experience on similar systems suggests that this method is preferred or required, or unless there is another compelling reason. In fact, despite the disadvantages of this method, it is very frequently applied on computer motherboards.

To see the disadvantages and merits of this method, let us first consider an analogous method used primarily for lower-frequency designs (from the good old days) and still applied very successfully in systems containing primarily analog and audio circuits. In this technique, called *star grounding*, the 0-V references are tied to earth ground (chassis) at only one point in the entire system. This is similar to inductive isolation of reference plane regions in the fact that it minimizes the propagation of noise by the use of an impedance that isolates the noise from the larger ground structure. For star grounding the high impedance is an open circuit, due to the absence of an otherwise local ground connection, whereas for inductive isolation an inductive component is used for the impedance. In principle, the two techniques have a few common traits.

When distributed components of a system are tied together in a star-grounded fashion, the ground of each component extends all the way back to one central ground and has no local alternative path to chassis ground. In Figure 10.19, two conditions are illustrated, one with the switch open and one with the switch closed. Components A and B can be different systems in different chassis, or different subsystems within a system. When the switch is closed, component B has a local dc connection to ground. When the switch is open, component B has no local connection to ground but still has some local parasitic capacitance to ground. The latter condition (switch open) is in the star-grounded configuration. At low frequencies, the star-grounded configuration has the advantage of having no extra ground loop since the potentially large loop between the local ground connections has been cut. Due to this, differential-mode emissions will be reduced. The star-grounded condition also has the advantage of having only one 0-V reference voltage and is immune to differences in the local 0-V

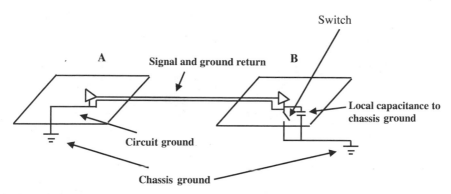

FIGURE 10.19   Star grounding. If switch is open, the systems are star grounded.

potential and induced differences from the large loop that would otherwise be present. All this works fine at low frequencies. However, recall that the differential emission envelope (Figure 10.5), decreases above a certain frequency for large loop size. Thus, the large loop that was avoided by using star grounding will radiate in the differential mode primarily at lower frequencies, which are not usually the most difficult emission problem in a high-speed digital system. Furthermore, for smaller systems in which the ground loop avoided by star grounding may not be so large as to be purely a low-frequency emitter, the capacitance to chassis ground at remote points (component *B*) can contribute to a resonance that can create an efficient radiator at high frequency. Even with no resonance, the capacitance closes the loop anyway, leaving no benefit of star grounding at high frequency. Finally, a distant ground, such as necessary for star grounding, will quite likely make the power delivery impedance too large for high-frequency current distribution. Thus, although star grounding has several advantages at low frequency, it should generally not be done on a high-speed digital system unless there is a compelling reason. A motherboard, for instance, should have many connections to chassis ground and should have low-impedance ground and power paths throughout. This is true particularly near high-frequency regions because without the dc path to chassis ground, a capacitive path to chassis may still be found and may excite a resonance. To summarize, in high-speed systems, avoid star grounding.

The emission effects of star grounding can be seen graphically in Figure 10.20. At low frequencies it can be seen that emission is reduced significantly; however, at chassis resonances, there may be large peaks. Not evident in Figure 10.20 is the fact that star grounding could also increase the impedance in the power delivery path and cause large common-mode voltages due to voltage drops in the power delivery system. This could negate any emissive gains as well as ruin signal integrity.

Consider Figure 10.11 again. The condition created is similar in many ways to that of star grounding. For instance, the power delivery is high impedance at high frequency (although this is an unintentional effect in star grounding).

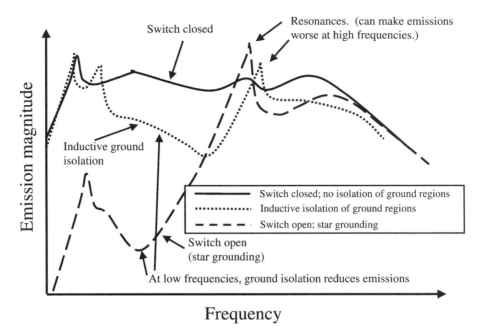

FIGURE 10.20   Effect of star grounding.

Also, the differential loop return has been disturbed with an impedance similar to star grounding in which the larger return loop was cut (although one may argue that it has instead caused any deviated current to return in a more labyrinthine fashion, thus increasing the loop area). Finally, similar to star grounding, there is a possibility of a capacitive return path that completes the cut loop for high frequency.

We have seen in Section 10.3.1 that high-frequency decoupling (above a few hundred megahertz) can be ineffective in a typical high-speed system. Thus, it is no surprise that the smaller power and ground regions in Figure 10.11 may not be adequately decoupled. The lack of adequate decoupling will create a noisy environment for the isolated component, which will have a negative effect on signal integrity and may cause false logic triggers. Furthermore, as already mentioned, if the ground and power planes are laid out such that the signal traces cross the gap in reference, more signal integrity effects will be introduced.

So, with all the negative effects of reference plane isolation, there exists the inescapable fact that it is sometimes the only feasible way to contain high-frequency noise that is causing a common-mode emission failure. In one case known to the author, a relatively strong memory clock buffer chip with 18 simultaneously switching outputs caused emission failures until the ground and power planes around the device were isolated and choked off with onboard ferrites. This particular example was very similar to Figure 10.11. In this case

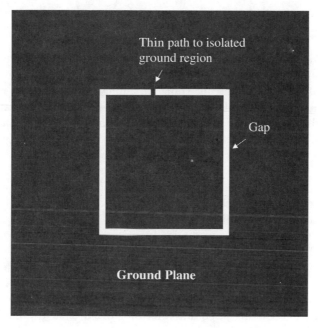

FIGURE 10.21 Routed ground isolation: does not work to reduce emissions.

all the outputs were 3.3-V 100-MHz memory clock outputs. The slowest rise and fall times for this particular device were specified to be about 1 ns. In retrospect, this problem should have been identified in the early architecture stages and should have been bypassed before it existed. As a general rule, involve EMI engineers early in the design stage.

Several references suggest, as a way to reduce emissions, slowing the edge rate of the elements in the system that is causing the noise. It is certainly true that slowing the edge rates will result in reduced emissions, since slowing the edge rate will result in fewer high-frequency components. However, there are many other variables inherent to a driver that may be of greater consequence. Furthermore, there is the fact that slowing the edge rates necessarily has a detrimental effect by increasing timing uncertainty in the system. Thus, slowing the edge rates should be done only as a last resort.

As a final note, measurements have shown that cutting ground or power planes in the manner shown in Figure 10.21 has a minimal effect in containing high-frequency noise. No matter how "thin" the path to the "island," the inductance is not effective.

10.3.3. Low-Frequency Decoupling and Ground Isolation

As a rule of thumb, frequencies below 30-MHz noise will be transmitted primarily by conduction, and noise frequencies above 30 MHz will radiate.

This is the reason that the radiated emission standards begin at 30 MHz, and below that frequency the compliance standards are considered conductive emissions.

It has already been noted that for low frequencies, star grounding has several benefits in minimizing noise. Thus the reader may have surmised that at low frequencies there is some advantage to be gained by isolating the ground and/or power planes of a low-frequency noise generator in order to contain the low-frequency noise. This is exactly the technique that will be advanced here. However, this method should be used only when necessary in a high-frequency system, and the trade-offs should be well understood.

In a case where low-frequency noise is causing a functional problem, or less likely an emission problem, the ground and power of the noise source can be isolated and effectively star grounded, unlike the rest of the system. Since high-speed systems typically have many ground paths to minimize ground impedance, low-frequency noise propagates very well through all dc paths in the system. To contain low-frequency noise, you simply cut the dc path. The only problem occurs when this affects the low impedance required for high-speed reference. Consider, for example, a low-frequency power source that dumps large amounts of current at a low frequency. The noise may already have its own separate power plane. The question confronting the designer is whether to cut the ground plane underneath the noise source to prevent noise from propagating to the rest of the system. Standard high-speed design practice says no. The standard design practice should certainly be adhered to if any high-frequency sources require a ground reference near the area to be cut. However, even if the cut is far away from any high-frequency reference point, cutting the ground decreases the total capacitance between the planes. The considerations contained in Section 10.3.1 regarding high-frequency decoupling should convince the reader that it is generally not preferable to cut the ground plane. However, there may exist cases in which the rule must be violated and the ground plane cut. There is some data to suggest that low-frequency noise present in a system has an ill but elusive effect on system clocking. It is possible that low-frequency noise can interfere with the operation of phased-locked loops (PLLs), which are used for frequency stability in clock drivers and receivers. This noise can cause the PLL to modulate at the same frequency as the noise. At noise frequencies sufficiently high, the noise will be beyond the PLL bandwidth and there will be no problem. At noise frequencies sufficiently low, all the PLL in the system will respond to the noise and copies of the clock reference will modulate in a fairly similar manner. The worst-case condition is when the frequency is between a few hundred kilohertz and about 5 MHz, at which condition some of the PLLs in the system will modulate and some will not. This noncorrelated modulation of the clock reference can cause large variations in the timing reference. Such an effect is very difficult to measure.

The final judgment is left to the designer. However, if the ground plane is cut, every precaution to minimize the area of the cut and the locality to high-frequency references should be made.

10.4. ADDITIONAL PCB DESIGN CRITERIA, PACKAGE CONSIDERATIONS, AND PIN-OUTS

Many of the aspects of board design for minimum emissions have already been mentioned. The benefits of low-impedance ground and power has been mentioned, along with the fact that there should be many chassis ground points both to prevent resonances from the capacitance to chassis and to further reduce the ground impedance. The value of maximizing the inherent power/ground plane capacitance has also been mentioned. Furthermore, it has been explained that traces should not cross gaps in the reference planes, particularly high-speed traces such as the system clock. Finally, the difficulties of high-frequency decoupling have been explored. In this section we detail some additional criteria for minimization of emissions and system noise. It has been assumed that in all cases, boards with sufficient numbers of layers to have at least one fairly continuous plane for ground and one plane for power have been used. Furthermore, it has been assumed that all boards under consideration are of controlled impedance design (i.e., 50-Ω traces) and have been implemented with some sort of matched termination. The details for boards with 'routed' power and ground are not covered in this chapter.

10.4.1. Placement of High-Speed Components and Traces

High-speed components should be placed ideally near the center of a motherboard, or at least far from the edge. Also, if a trace is routed near a vacancy in the ground plane, for instance, fringing fields, which would otherwise be referenced to ground, can radiate to space. These fringing fields can also capacitively couple to the chassis or other structures and excite a resonance. Similar arguments apply to vias in close proximity to high-speed traces and components. Vias cut the power and ground planes, resulting in a discontinuous plane. Their placement should be considered carefully.

10.4.2. Crosstalk

Crosstalk is typically considered as a signal integrity parameter, but it is also a parameter that affects the mobility of noise in a system. More crosstalk results in a higher probability that noise will find an efficient radiator, such as crosstalk to a trace that ultimately is routed to an external cable. Basically, by limiting crosstalk the number of paths that high-frequency components may take is limited, thus limiting the probability that a radiation mechanism will

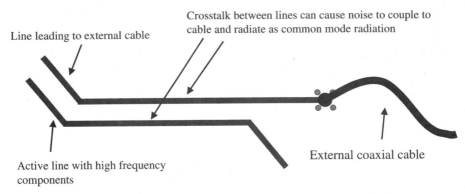

FIGURE 10.22   Crosstalk to line feeding external cable.

be excited. Although it is difficult to make statistical judgments, the designer should realize that less crosstalk will generally result in less radiated emissions. Consider a trace that feeds an external cable adjacent to another board trace, as shown in Figure 10.22. Crosstalk can directly create common-mode noise on the trace feeding the cable and could potentially cause large emission problems.

Crosstalk falls off with conductor spacing exponentially. When the minimum spacing is being chosen, early in the design cycle, the crosstalk versus distance should be plotted. If the design is in the region where a small amount of increase in trace-to-trace distance will result in a dramatic decrease in crosstalk, increasing the spacing should be considered. Other variables that affect crosstalk, such as distance from ground plane, should be plotted in the same manner. Basically, if the crosstalk is in a "steep" portion of the graph, a little extra spacing should be an easy sell to the design team. Furthermore, target or known-sensitive traces can be picked, and the crosstalk on these traces can be minimized.

10.4.3. Pin Assignments and Package Choice

The pin assignment recommendations and package choice criteria of Chapter 5 should be adhered to. The importance of good power and ground pin selection for emissions compliance has to do with minimizing the impedance of power delivery to the silicon, minimizing the area of return path loops, and increasing the frequencies of harmful resonance effects. When possible, outputs of different frequencies should not share adjacent power or ground pins (and should not be placed next to each other). Furthermore, the pin-out should be chosen with decoupling in mind. As described in Chapter 5, it is usually optimal to place power and ground pins adjacent to each other to increase the coupling between power and ground. In a case where it is known in advance that there will be a decoupling impossibility, the use of specialty low-inductance capacitors may be considered. This may have an effect on the pin-out chosen.

10.5. ENCLOSURE (CHASSIS) CONSIDERATIONS

In this section we detail parameters of the box that surrounds the system or components therein.

10.5.1. Shielding Basics

Two basic mechanisms of shielding exist: *reflection* and *absorption*. These concepts, when applied to radiation, are not fundamentally different than reflection in transmission line systems. It is basically a matter of the wave impedance and the impedance of the shield. In the far field, where the wave impedance is 377 Ω, a conductor will look like a short circuit and reflect the wave much as it would do in a transmission line. Similarly, if a conductive shield is in the presence of a low-impedance radiator (current loop) in the near field, the shielding may not be effective because the wave impedance is low and the conductive shield looks like much less of an impedance discontinuity. Thus, the shielding effectiveness can be lower in the near field. However, the shield effectiveness can also be higher in the near field if the radiator is a high-impedance (voltage) radiator and thus may have a near-field wave impedance greater then 377 Ω. The equations in Section 10.2 are applicable here to determine wave impedance and near and far-field conditions.

Another parameter of interest when considering a conductive shield is the skin depth. The reader will recall that the penetration distance of a field into a conductor decays exponentially, $e^{-x/\delta}$, where x is the distance beneath the surface of the conductor and $\delta = \sqrt{\rho/\pi F \mu}$ is the skin depth, which is the distance at which the field has decayed to $e^{-1} = 37\% = -8.7$ dB of the strength at the surface. Figure 10.23 shows a field impinging on a shield. There is a reflection at the impedance discontinuity of the surface of the conductor and the opposite surface. The exponential decrease in field magnitude inside the conductor is also shown.

Thus, depending on the thickness of the conductive shield, measured in skin depths, the amount of the field that penetrates the shield can be calculated. The shield effectiveness (SE) due to penetration offered for a sheet metal shield can be calculated as

$$\mathrm{SE}\alpha_{\text{sheet metal dB}} = 20\log\frac{E_{\text{inc}}}{E_{\text{tunnel}}} = 20\log\frac{E_{\text{inc}}}{E_{\text{inc}}e^{-x/\delta}} = 20\log[e^{x/\delta}] = 20\left(\frac{\ln e^{x/\delta}}{\ln 10}\right)$$

$$= 20\left(\frac{x}{\delta\ln 10}\right) = 8.7x\sqrt{\frac{\pi F \mu}{\rho}} = 0.0172x\sqrt{\frac{F\mu_r}{\rho}} \qquad \text{decibels}$$

$$(10.20)$$

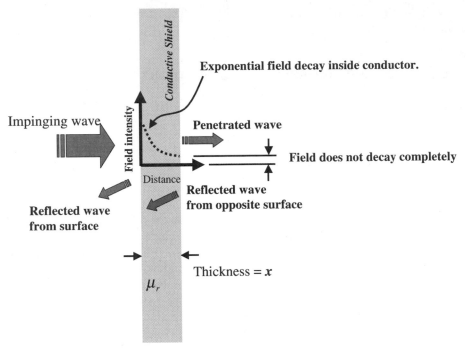

FIGURE 10.23   Skin depth as a shielding mechanism.

where x is the shield thickness, in meters; ρ the resistivity of the shield, in ohm meters; μ_r the relative permeability of the shield material; δ the skin depth of the material; and F the frequency of the wave.

From Table 4.1 the resistivity of copper is $\rho_{Cu} = 1.72 \times 10^{-8} \, \Omega \cdot m$, whereas pure iron is $\rho_{Fe} = 10 \times 10^{-8} \, \Omega \cdot m$ and steel ranges from 1.2 to 120 times the resistivity of pure iron. The relative permeability of iron and/or steel ranges from 110 to 14,800; however, ranges from 300 to 1000 are typically used for shielding calculations. Thus, although pure copper is more conductive (less resistive) then iron-based metals, the iron-based metals allow less penetration (recall that increased permeability decreases the skin depth). This is important except for conditions where the shield is thin ($x \ll \delta$). When the shield is thin, most of the shielding is accomplished due to reflection, and a higher-conductivity metal such as copper would be an advantage.

For materials other then sheet metal, such as conductor impregnated plastic, equation (10.20) is generalized as

$$SE\alpha = 0.0172xq\sqrt{\frac{F\mu_r}{\rho}} \qquad \text{decibels} \qquad (10.21)$$

where q is a constant that varies with the conductive material or coating and x is the thickness. For conductive surface coatings, q will vary from 0.2 to 1.0 [White, 1986].

Reflection is, of course, the other parameter that determines shield effectiveness (SE). Shield effectiveness due to reflection is defined as

$$SE\rho_{dB} = 20\log \frac{(1 + Z_{wave}/Z_{shield})^2}{4(Z_{wave}/Z_{shield})} \qquad \text{decibels} \qquad (10.22)$$

where Z_{shield} is the per square impedance of the shield (the same as the equivalent metal surface impedance in ohms per square for sheet metal, or this number is provided in data sheets for special shield materials), and Z_{wave} is the wave impedance at the distance of the shield. Again, the shield effectiveness is a measure of the decrease in radiated emissions [White, 1986].

For wave impedance values much higher than the shield impedance, as is usually the case for conductive shields, equation (10.22) reduces to

$$SE\rho_{dB} \approx 20\log \frac{Z_{wave}}{4Z_{shield}} \qquad \text{decibels} \qquad (10.23)$$

where the wave impedance is 377 Ω in the far field or can be derived for the near field from equation (10.8) or equation (10.3), depending on the nature of the radiator. One can see that for low-impedance waves, such as in the near field of a low-impedance loop radiator, it may be difficult to get reflective performance from the shield.

Notice that unlike penetration performance, reflective performance from thin shields is not a function of frequency. Thus, thin conductive shields can be implemented successfully using primarily reflection as a shielding mechanism. The condition to avoid with such a shield is, of course, near-field low-impedance conditions. Conductor-coated, conductor-impregnated, or conductive plastics may also be used. These materials exist in several varieties, including particle-loaded plastic, fiber-impregnated plastic, carbon plastics, and plastic-coated screen. The type can be chosen for the application in mind. For instance, due to the larger contact area between the conductive elements, fiber-loaded plastic can be had in a lower mass density than can particle-loaded plastics. Wear performance is another criterion. The reflective performance, however, is determined by the surface impedance Z_{shield}.

The total shield effectiveness is calculated as

$$SE_{shield} = SE\alpha + SE\rho \qquad (10.24)$$

10.5.2. Apertures

The shielding effect of an enclosure is only as good as its weakest link. It is no surprise that holes in the shield will often be the dominant variable in shielding effectiveness. Fortunately, there are techniques available to predict and to minimize the radiation from apertures in the shield. Some common reasons to have apertures are for viewing screens, thermal airflow, switch holes, cable holes, and mating seams, to name a few. In all cases, the amount of compromise of the shielding can reasonably be predicted, at least for worst-case values. Typically, during the design stage, the dominant apertures (based on judgment) should be calculated in order to have a rough idea of the amount of shielding to count on. Calculating the worst case of every aperture in the design stage could lead to shield overdesign.

Rectangular and Circular Apertures. Consider the rectangular aperture shown in Figure 10.24. An impinging wave on the surface containing the aperture, and polarized such that the electric field is 90° to the direction of the slot, produces the worst-case noise transmission. In this case the opposite surfaces of the slot will look like a capacitor, and because current will have to flow around the slot, there will be an inductive component as well. Thus, much like a transmission line, the slot will look like an *LC* circuit and will have an impedance that depends on the slot dimensions. In fact, the panel with the rectangular hole can be thought of as a slot transmission line that is shorted on both ends, as suggested by the cross section of the panel side shown in Figure 10.24. The ratio of the slot impedance and wave impedance will determine the magnitude of the reflection coefficient and subsequently the shielding effectiveness. Equation (10.25) is a practical formula for computing the shield effectiveness of a slot [White, 1986]. Note that equation (10.25) must always be less than or equal to the shield effectiveness without a slot, as calculated in equation (10.24).

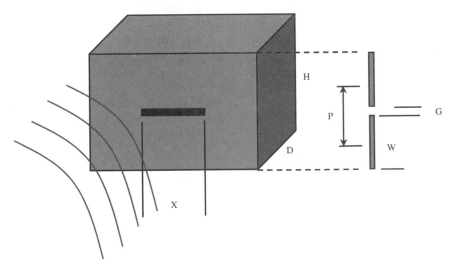

FIGURE 10.24   Chassis with a hole in it.

$$SE_{slot} = 157 - 10\log(XGF^2) + SE_{shadow} + SE_{WGBCO} \quad dB \quad (SE_{slot} < SE_{no\ slot})$$

$$(10.25)$$

where SE_{slot} is the shield effectiveness of the slot, in decibels. SE_{shadow} is the shielding performance due to the shadowing effect, depending on the location of the slot, the geometry of the enclosure, and the location of the radiator. This is explained below. SE_{WGBCO} is the effect of the "deepness" of the slot. Deep slots will radiate less due to an effect known as *waveguide beyond cutoff*. This effect is also explained below. X and G are the length and width in meters of the slot, as shown in Figure 10.24. F is the frequency under consideration, in hertz. The terms to the right of 157 dB constitute the effective slot leakage.

Equation (10.25) is also valid for a circular aperture if 2 dB is added to the 157 term. This comes from the fact that the area of a circle fitting inside the square made when $X = G$ is 0.785 times the area of the square made by XG. $20\log 0.785 = -2$ dB less area, and since less slot area results in more slot attenuation, the 2 dB should be added. It should also be noted that equation (10.25) is for the worst case, where the electric field is perpendicular to the slot. For cases where this is not true, the shield can be far more effective.

Equation (10.25) cannot be more effective than the box without the aperture, and cannot be lower (less effective) than

$$SE_{slot} = 6\log\frac{X}{G} + SE_{shadow} + SE_{WGBCO} \quad (10.26)$$

Shadow Effect. The shadow effect noted in equation (10.25) results from the fact that not all radiators inside the box are positioned well to radiate toward the slot. This is affected by the aspect ratios of the box relative to the slot position and the frequency of the radiation under consideration, and of course it is also affected by the slot size. To visualize the effect of shadowing, imagine shining a flashlight into a slot from outside the box. Where the light falls inside the box is the worst site to place a radiator. In reality, the flashlight example is not quite true, however, because electromagnetic waves of typical emission frequency passing through a slot will fringe and exhibit lobes.

The behavior of the radiation after it enters through the slot from outside the box will depend on its wavelength relative to the length of the slot. At low frequencies, where $X < \lambda/2$ (or, in other words, $F < c/2X$, where X is the slot length and c is the speed of light), the wave will tend to stay within 120° radial from the slot. Thus, at low frequencies, radiators positioned inside the box at angles greater than 120° from the slot will have a significant shadow effect. For positions at angles less than 120°, the shadowing effect will be minimal. This agrees with intuition, which says that the inside surface of the shield containing the slot is a well-shadowed surface (180° away) and directly opposite the slot (90° away) is not a well-shadowed surface.

At higher frequencies, the radiation pattern inside the box resulting from an impinging waveform outside the box will develop lobes of maximum and

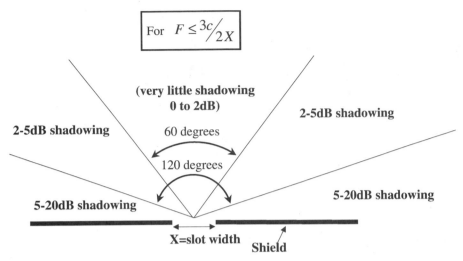

FIGURE 10.25 Shadowing effect.

minimum radiation intensity. The higher the frequency, the more lobes will exist. The lobes tend to decrease the shadow effect because they will bend around the corners. Above a frequency where $X = 2\lambda$ ($F > 2c/X$), no shadowing effect should be considered.

The numbers shown in Figure 10.25 can be used as an estimate for shadow effect for $X \leq 3\lambda/2$ (or $F \leq 3c/2X$). For more precise estimates of the shadowing effect, other references provide tabulated data based on volume integration of the box volume in which the shadow effect can be looked up using the box dimensions and slot dimensions [White, 1986]. It is intuitive that shallow boxes can produce significantly more shadow effect than can a deep box. Similarly, small slots have significant shadow effects.

Waveguide Beyond Cutoff Effect. When an aperture has a significant depth, it will begin to exhibit the properties of a waveguide. For frequencies less then the frequency necessary to excite the propagating modes of the waveguide, the depth will improve the shield effectiveness of the aperture. This is basically operating the waveguide below its operating frequency. This is known as the *waveguide beyond cutoff effect*. Waveguides have several propagation modes of operation. If a frequency in a waveguide is below its dominant operating frequency, it is said to be in *cutoff*. Below the cutoff frequency, the radiation will be attenuated. Above the cutoff frequency, the radiation will pass relatively unhindered.

There are multiple cutoff frequencies for rectangular waveguides, depending on the mode of propagation being considered (derivations of these cutoff frequencies can be found in several electromagnetic textbooks). However, as an approximation, a dominant cutoff frequency can be calculated for a given aperture from the equation

$$F_{\text{cutoff}} \approx \frac{c}{6X}$$ (10.27)

where X is the length dimension of the aperture and c is the speed of light. Note that this in only an approximation to get a handle on the problem.

The effective shielding effect for frequencies below the cutoff frequency is approximated by

$$SE_{\text{WGBCO}} = 30\left(\frac{t}{X}\right) \qquad \text{decibels}$$ (10.28)

Furthermore, equation (10.28) applies generally only when the thickness of the shield, t, has a significant depth compared to the slot dimension; thus $t/X > 2$. Above the cutoff frequency, there will be no shielding due to this effect. Thus, one way to increase the total shielding effectiveness [equation (10.25)] when apertures are necessary is to use thick shielding material or deep slots.

Modern chassis designs are usually made of thin sheet metal, so this effect is negligible. Obviously, to use thicker metal would dramatically increase the weight and cost of the system. However, the innovative engineer may take advantage of the effect by using some kind of lightweight shielding material or by increasing the thickness of the shield only in the vicinity of apertures.

Screen Mesh. The shielding offered by a screen mesh is calculated as [White, 1986]

$$SE_{\text{screen}} = 164 - 20\log(SF)$$ (10.29)

where S is the spacing between the wires in the mesh. Note that this equation is similar to (10.25).

Seams. Seams where chassis surfaces come together can be a source of leakage. The length of the seam is well defined; however, the width of the gap can vary widely. Judgment can be applied to estimate the gap width or the gap width can be measured with a feeler gauge. Then the equations for slot apertures can be applied. If a conductive gasket is used, the leakage can be estimated from the vendor-supplied data.

Multiple Apertures. Because the aperture leakages calculated in the preceding sections give values in decibels, the effective leakage of several apertures must be placed in linear units before adding them for a combined effective leakage:

$$\text{worst-case effect of all apertures} = -20\log\sum\log^{-1}\frac{|L_{i\_\text{dB}}|}{20}$$ (10.30)

where L_{i_dB} is the leakage effect in decibels of the aperture i as included in equation (10.25) or a different appropriate equation. For example, if the leakage of an aperture is equal to $L_{hole} = -3$ dB, meaning that the hole degrades the shield effectiveness of the shield by 3 dB, four equal holes would degrade the shield by

$$20\log\left(4\log^{-1}\tfrac{3}{20}\right) = 15 \text{ dB}$$

For apertures that are identical to each other, (10.30) reduces to

$$L_i - 20\log N \tag{10.31}$$

where N is the number of apertures.

An important observation to make is that emissions are reduced if one aperture is broken up into smaller ones. To see this, consider equation (10.25). Note that the leakage is proportional to slot dimension. If a slot aperture is subdivided into N smaller apertures, the improvement in shield effectiveness is calculated as

$$SE_{\text{gained from subdividing}} = 10\log N \tag{10.32}$$

10.5.3. Resonances

Resonance conditions in the system can compromise an otherwise good design and can result in very efficient radiators. All the features of the entire final structure are candidates for resonance, and the resonance conditions are not typically predictable. This is one of the reasons that the possibility of changes late in the design cycle is always a threat. Unanticipated resonators could be heat sinks, chassis structures, resonant apertures, cables, traces, patches, or a wide variety of other structures. As a general rule, to avoid resonance, structures in the system should not be left floating. For instance, heat sinks should be grounded. The author knows of one incident where a system clock chip was located underneath a heat sink for a CPU. In this case the clock chip was coupling energy upward to the heat sink and causing a resonant condition. Increasing the distance between the heat sink and clock chip solved this problem. Although resonances are generally not predictable (i.e., the resonance of the heat sink did not seem to correlate to any particular dimension of the heat sink), it is very helpful to keep in mind that small structure changes can make big differences. A knack for applying trial and error is very helpful in this regard.

Resonant Slots. One predictable resonance is that of a resonant slot aperture. At a slot length of $\lambda/2$, the slot is resonant and can be considered a perfectly tuned dipole. This antenna will exhibit no shielding properties. This, along with the reasons discussed previously, is a good motive for keeping slot sizes small.

Losses. Chassis resonances often dominate high frequency emissions. If resonances dominate emissions, the importance of making the resonances lossy, which will decrease the effective Q, can be seen. Often, this can be accomplished by placing lossy material inside the enclosure. When experimenting with such materials, it should be realized that a populated PCB board is itself a lossy component. Experiments, for example, of chassis resonances in the absence of a populated board will yield results that may not be as significant when a populated board is present in the system. Subsequently, when experimenting with lossy materials in the chassis, make sure that the fully populated board is included.

10.6. SPREAD SPECTRUM CLOCKING

A technique has recently become popular in high-speed digital designs known as *spread spectrum clocking* (SSC). SSC derives its emission benefits from slowly modulating the frequency of the system clock back and forth a small amount. Since many events in the system are synchronized to the clock signal (i.e., CPU circuitry is synchronized to the clock signal), changing the frequency of the clock results in many radiation mechanisms shifting in frequency, not just direct radiation from the clock circuitry itself. Thus, by slowly shifting the frequency of the clock back and forth, peaks in the frequency spectrum of operation of the system will move back and forth in proportion to the magnitude of modulation. In a time average (which is how the products are measured for emission compliance), the energy of the peaks will be smeared out and result in lower peaks. Thus, due to this smearing effect, peak emissions at any particular frequency will be reduced accordingly and will subsequently pass the emissions standard even though the total emitted energy is not likely to change. This effect is shown in Figure 10.26.

The shape of the smeared spectral peak is determined by the modulation profile, that is, how the frequency varies between its modulation limits. It can be shown that one modulation profile that achieves a flat smeared spectrum, as shown in Figure 10.26, is shown in Figure 10.27, nicknamed the *Hershey Kiss profile*. Profiles other then this may have "rabbit ears" in the speared profile and thus not have as much peak reduction. The Hershey Kiss modulation profile is patented by Lexmark, however, and may involve some additional component cost (passed on from the clock vendor implementing this profile), due to royalties.

One problem with clock modulation is that it can induce timing skew in the system phase-locked loops (PLLs) that track the clock frequency in other parts of a computer system. In general, the slower the frequency changes, the less induced skew there will be downstream between all the PLLs. Thus, the Hershey Kiss profile is not optimal for tracking skew. Because both sinusoidal and

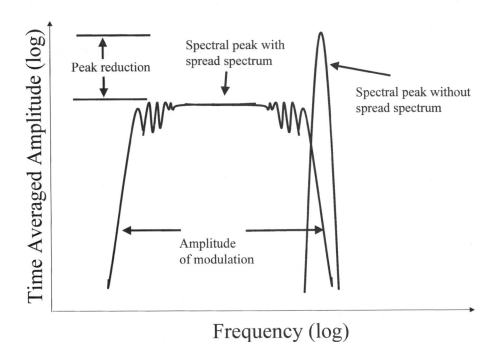

FIGURE 10.26 Spectral content of the emission before and after spread spectrum clocking.

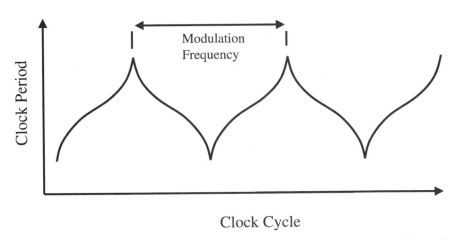

FIGURE 10.27 "Hersey Kiss": clock modulation profile that produces a flat emissions response.

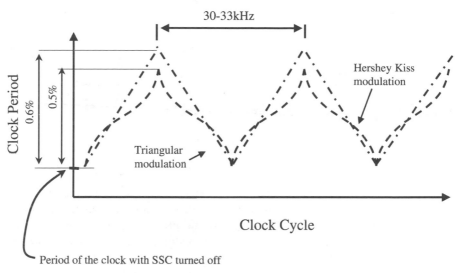

FIGURE 10.28 Spread spectrum clocking profiles.

triangular modulation involve slower changes of frequency, either would be better for tracking skew but would not result in the maximum peak reduction benefit gained by Hershey Kiss. However, due to the gain in tracking skew using a non–Hershey Kiss profile, a bit more amplitude frequency variation can be introduced, yielding roughly equivalent benefits. As a de facto industry standard, the SSC modulation, if implemented, should be 0.6% down-spread (meaning that the clock frequency is varied only downward from the nominal) assuming a triangular modulation profile, or 0.5% down-spread assuming a Hershey Kiss profile. Each profile is shown in Figure 10.28. The modulation frequency should be in the range 30 to 33 kHz (this modulation frequency is chosen to avoid audio-band demodulation somewhere in the system). Sinusoidal modulation profiles are generally not used because they exhibit a small spectral peak reduction. The reader should note that although SSC will decrease emissions, it will induce timing uncertainties into the system, especially in the phase-locked loops.

CHAPTER 11

High-Speed Measurement Techniques

So far, we have covered the fundamental aspects of designing a digital system, ranging from fundamental transmission line theory to design methodologies. After the design is finished and the prototype products have been received, it is necessary to validate the design. Design validation is necessary to prove that the design is robust, to ensure that no critical aspect of the design was neglected, and to verify that the simulations were accurate enough to predict system performance. Since each particular design is unique and requires specific validation methodologies and equipment based on its functionality and bandwidth, it is difficult to provide a general recipe that steps the engineer through validation. Subsequently, we will not concentrate on a general validation methodology, but instead, on a few aspects of high-speed measurements that will be useful in the validation of virtually any high-speed digital system.

In high-speed digital design and technology development, proper measurement techniques are critical to ensure accurate assessment of system timing margin during design validation and to verify that the simulations correlate with laboratory measurements. Measurements in many ways are the link between simulation and reality and are key to aiding the development process as a check and balance to ensure a robust design. As system speeds increase, the ease of capturing accurate repeatable measurements is becoming increasingly difficult to achieve. Often in modern high-speed designs, the setup and hold margins are only a few hundred picoseconds. To resolve this margin properly and verify the system timings, it is necessary to measure the system to a resolution of approximately one-tenth of the available margin. Subsequently, both the equipment and the measurement techniques must be sufficient or it will become impossible to fully characterize a design in the laboratory.

This chapter is intended to give the reader a good understanding of the basic types of equipment available and various measurement techniques that can be used to help extract the critical electrical characteristics for model correlation as well as system validation.

11.1. DIGITAL OSCILLOSCOPES

Today, the most prominent basic tool for electrical analysis is the digital oscilloscope. The instrument operates by waiting for a *trigger event* and then recording voltage and time information after the trigger event. Typically, the trigger event is a rising or falling voltage transition past a specified voltage. The instrument typically operates in real-time and/or equivalent-time mode. In *real-time operation*, the instrument takes consecutive samples of the voltage waveform after the trigger event and constructs the signal accordingly. If the sample rate is not sufficiently fast, the waveform will not be constructed accurately. The sample rate of the instrument strongly affects the quality of the reconstructed waveform in real-time operation. In *equivalent-time operation*, the instrument assumes that a periodic waveform is under measurement and constructs a displayed waveform from samples taken at varying delays from different triggering events. Basically, in this mode the waveform is constructed as a combination of different periods of the periodic waveform. Real-time and equivalent-time operation are explained further later in this chapter. The analog bandwidth of the instrument also affects the quality of the reconstructed waveform. The analog bandwidth gives information on how high frequencies are attenuated when entering the instrument. The bandwidth typically reported for an oscilloscope is the frequency at which the input will be attenuated by 3 dB. If the bandwidth of the oscilloscope is not sufficiently high to capture all significant frequency components of a signal, the measured waveform will not be constructed accurately.

11.1.1. Bandwidth

The input bandwidth to a scope is similar to an *RC* circuit, where the *R* and the *C* represent the input capacitance and impedance. Subsequently, as derived in Appendix C, the frequency spectrum of an edge passing through an *RC* circuit is related to the subsequent rise or fall time by equation (11.1). The rise time in equation (11.1) is the fastest rise time that can be passed without distortion, assuming the corresponding bandwidth. Since spectral content and rise or fall time are related to each other as shown in equation (11.1), it is easy to see that the input bandwidth of the scope can filter out high-frequency components of the signal rise time and degrade the signal edge. For example, if the system rise time is 100 ps but the bandwidth of the probe is 2.5 GHz and the input to the scope is 1.5 GHz, the rise time constructed by the oscilloscope will be 290 ps as calculated using the root mean square, as shown in equation (11.2). The terms T_{scope} and T_{probe} in equation (11.2) are the rise and fall times calculated with equation (11.1) from the respective bandwidths of the scope and probe. Equation (11.1) is valid for rise and fall times measured between 10 and 90% of the amplitude. For different definitions of the rise and fall times (i.e., 20 to 80%), different equations will result.

$$\text{bandwidth}_{3 \text{ dB}} \approx \frac{0.35}{T_{\text{rise/fall}}} \tag{11.1}$$

$$T_{\text{measured}} = \sqrt{T_{\text{scope}}^2 + T_{\text{probe}}^2 + T_{\text{signal}}^2} \tag{11.2}$$

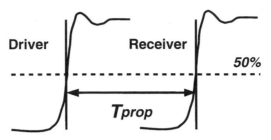

FIGURE 11.1   Illustrative example of flight time (T_{prop}) showing flight time as function of voltage threshold.

where $T_{measured}$ is the rise or fall time measured by the scope, T_{scope} and T_{probe} are the minimum rise and fall times that the input of the scope or probe will pass as solved by equation (11.1), and T_{signal} is the actual edge rate.

As illustrated in Figure 11.1, the time interval measurement (i.e., flight time) is a function of voltage because the measurement is taken at or near the threshold voltage of the receiving buffers (for this case the 50% amplitude point). The key item to understand is that if the scope degrades the signal rise time significantly, flight times and flight-time skews will not be characterized accurately. For example, if two signals exhibit a 100-ps variation in threshold crossing that is caused by a difference in edge rates and the scope used to measure the variation exhibits a bandwidth that is not sufficient to resolve this edge difference, the measured skew would be less. This is demonstrated in Figure 11.2. It is important that the input bandwidth of the oscilloscope and the probe be sufficient to measure the edge rates and the skews to sufficient resolution. Another source of measurement error due to inadequate scope bandwidth is the loss of high-frequency aberrations which will be attenuated. It is important to note that sampling, discussed next, will also play a critical role.

11.1.2. Sampling

A critical parameter in scope performance is sampling method and sample rate. *Sampling* refers to how the scope captures the data being measured. Digital scopes measure voltages at specific time intervals. These measurements are then reconstructed on the screen. If the time interval between voltage samples is small enough, the result will be an accurate representation of the waveform. *Sampling rate* refers to the time interval between samples. For example, if a scope has a sampling rate of 10 gigasamples per second, it means that a sample will be taken every 100 ps.

General oscilloscopes come in two different types of sampling, real time and equivalent time. Most high-speed designs require real-time sampling for capture of nonperiodic or nonrepetitive (random) events. *Real-time sampling*

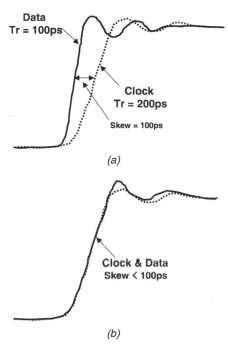

FIGURE 11.2 Example of oscilloscope bandwidth limiting resulting in an error of measured clock to data skew. (*a*) actual signals; (*b*) measured on oscilloscope.

refers to a signal that is captured by reconstructing the sample points over a single time interval. That is, the scope will take consecutive voltage samples, following a trigger event, and connect the dots to reconstruct the waveform. The samples used to construct the waveform are taken and displayed in the same sequence as they actually occurred on the signal, hence the name: real-time. If the sample rate is sufficient, the waveform will be represented properly. However, if it is not, the signal shape will not reflect reality. Figure 11.3*a* depicts the effect of an inadequate sample rate in real-time mode.

When real-time sampling is not sufficient to capture the waveform, equivalent-time sampling can sometimes be used. *Equivalent-time sampling* generates the waveform by piecing together samples from different periods of a periodic waveform. The sample points are taken at slightly different time intervals so that enough points can be sampled to reconstruct the waveform adequately. The input signal is required to be periodic for equivalent-time sampling. The equivalent sample time interval is much smaller allowing for potentially much more accurate reconstruction of periodic waveforms. To accomplish this task, a delayed trigger is used in either sequential or random order. For sequential sampling the trigger delay is a fixed amount. This reconstructs the waveform by capturing different points along the waveform in time until enough points

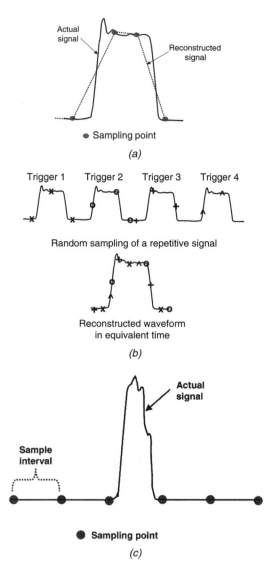

FIGURE 11.3 (*a*) Inadequate real-time sampling. (*b*) A random delayed trigger of a repetitive signal can be used to reconstruct the waveform in equivalent time. (*c*) A voltage glitch can be missed altogether if the maximum sampling rate is wider than the glitch.

are stored to reproduce the waveform accurately. Random repetitive sampling is similar to sequential except that sampling is completed randomly versus a fixed delay trigger. An example of equivalent-time sampling is illustrated in Figure 11.3*b*.

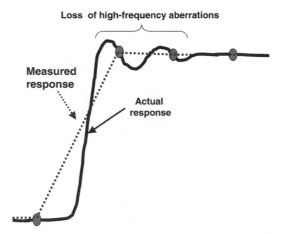

FIGURE 11.4   How measurement resolution loss occurs when the sampling rate is inadequate.

It should be noted that equivalent-time sampling is only sufficient to capture repetitive events. To capture random or nonperiodic events, it is necessary to use real-time sampling. This causes a problem in high-speed measurements. Equivalent time will average out nonperiodic events, and real time may not capture them if the sample rate is not high enough. As illustrated in Figure 11.3c for nonperiodic waveform capture, the real-time sampling rate can determine whether a glitch is captured at all. Random glitches in the signals can cause system failures. If the sample rate is not sufficient, it becomes extremely difficult to capture these random events. In cases where the width of the glitch is less than the sampling rate, signal capture may be missed altogether. Similarly, in Figure 11.4, the actual placement of the edge can be unknown if a sample does not fall on the transition. Figure 11.5 shows another example of lost information due to inadequate sample rate.

11.1.3. Other Effects

There are many other effects of the measurement equipment that can affect measurement accuracy. Although a detailed account of each of these effects is beyond the scope of this book, some of them are mentioned here briefly.

Internal Oscillator Accuracy and Stability.   The timing reference points inside the oscilloscope are typically based off a crystal oscillator. The quality of the crystal chosen by the manufacturer will affect the accuracy and frequency stability of the instrument. The oscillator will have a nominal accuracy and will also change slightly with time. Oscilloscopes tend to be far more accurate with relative time measurements (such as comparing the edge rates of similar signals) than with absolute time measurements (such as measuring a long pe-

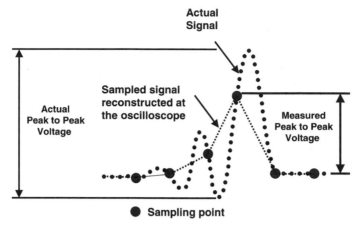

FIGURE 11.5 Vertical resolution loss due to inadequate sampling of high-frequency signal.

riod very accurately). With relative time measurements, the nominal accuracy may not matter. However, since the frequency accuracy of the internal oscillator can vary with time, relative measurements separated over long lengths of time may yield inaccurate results. The oscillator may also have temperature sensitivities and other imperfections.

Analog-to-Digital Converter Resolution. The voltage amplitude has a finite resolution due to the analog-to-digital converter used internally. Most oscilloscopes use an 8-bit analog-to-digital converter which allows a granularity of 256 levels of amplitude. Typically, the 256 levels are scaled to the display. Thus, at any time the fraction of the display used is the fraction of the analog-to-digital converter that is being used. Thus, for accurate amplitude measurements, the waveform should be scaled to fill a good fraction of the display.

Trigger Jitter. This is the variation in placement of the trigger by the instrument. This can lead to apparent time variation of a measurement.

Interpolation. How the instrument chooses to "connect the dots" can affect the shape of a reconstructed waveform. Various interpolation algorithms exist and are usually selectable. The user may also choose to shut off interpolation and see only the captured points.

Aliasing. In certain measurements, effects can be observed which are a result of sampling a frequency or frequency component at a sampling rate less than required to capture the signal. This can sometimes result in apparent lower-frequency components that do not actually exist in the signal. Mathematically,

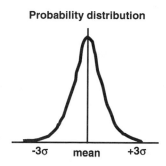

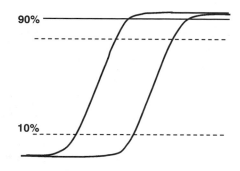

FIGURE 11.6 The probability distribution can be used to incorporate the effects of jitter on rise time.

to capture a frequency component, you must sample at a minimum rate of twice the frequency of interest. For practical instruments, it is typically assumed that you must sample at four times the maximum frequency of interest. This is why the maximum sample rate of an oscilloscope that is to be used for real-time sampling is typically four times its rated analog bandwidth. Aliasing is most pronounced at sample rates less than maximum.

11.1.4. Statistics

Statistics are necessary to indicate measurement repeatability and reproducibility for systems that include significant jitter, which is variation in a measurement. Jitter is so named because when observed on the screen the waveform will appear to shake or blur if jitter is present. By utilizing some basic statistical parameters, one can check to see if the particular measurement is stable. Measurement stability can be verified by taking a number of samples and computing the mean and standard deviation from the Gaussian distribution. This is a function on many modern oscilloscopes. This should be completed on any new measurement setup to verify that measured differences in the results are not within the noise of the measurement. Figure 11.6 illustrates an example of a Gaussian distribution of a rise-time measurement. Because jitter exists in the system, the edge rate may be changing, which could make it difficult to produce repeatable results. By measuring the mean and standard deviation for a parameter such as rise time, one can compute more accurate and repeatable results in the presence of jitter effects by computing the rise time from the statistical means at the two measurement points (i.e., 10% and 90%).

11.2. TIME-DOMAIN REFLECTOMETRY

In previous chapters we saw how the waveform at a driver is determined by reflected components. Time-domain reflectometery (TDR) refers to measuring a system by injecting a signal and observing the reflections. Typically, a step function with a defined transition time is driven into a transmission line and the waveform at the driver is observed. Thus, the features of all the downstream reflection effects on the driver end of the line, such as ledges, overshoot, and undershoot, can be observed and back-calculated to characterize the impedance discontinuities on the transmission line. A similar method which looks at transmitted waves rather then reflected waves, called a TDT (time-domain transmission), is explained later in the chapter.

At high speeds, the interconnect can be the major factor in limiting system performance. To predict performance early in the design phase, it is extremely important to characterize and model the interconnect accurately. Time-domain reflectometry (TDR) measurements are common for PCB board characterization. The key advantage of the TDR over frequency-domain measurements is the ability to extract electrical data relevant to digital systems, which represent time-domain signals. As mentioned in previous chapters, digital signals represent wide bandwidth signals, not single frequencies as prominent in typical microwave designs. Extracting data in the time domain provides voltage–time data that relate back to system operation. By utilizing the TDR, one can extract impedance, velocity, and mutual and self-transmission line electrical parameters. In the following section we discuss the basic theory and operation of the TDR to extract interconnect electrical characteristics.

11.2.1. TDR Theory

The TDR uses basic transmission line theory by measuring the reflection of an unknown device relative to a known standard impedance. This is based on knowing that energy transmitted through any discontinuity will result in energy being reflected back. The reflection magnitude will be a function of both the transmitted energy and the magnitude of the impedance discontinuity. The time delay between transmitted and reflected energy will be a function of distance and propagation velocity.

Impedance information is extracted by calculating the reflection coefficient, ρ, as a function of the incident and reflected voltages. Using the TDR reference characteristic impedance and ρ, the trace characteristic impedance can be computed using the equations

$$Z_{\mathrm{DUT}} = Z_o \frac{1 + \rho}{1 - \rho} \tag{11.3}$$

$$\rho = \frac{V_{\mathrm{reflected}}}{V_{\mathrm{incident}}} = \frac{Z_{\mathrm{DUT}} - Z_o}{Z_{\mathrm{DUT}} + Z_o} \tag{11.4}$$

The equations above can be used to calculate the trace impedance (Z_{DUT}) [device under test (DUT)], where Z_o represents the output impedance of the TDR. The oscilloscope will calculate the value of ρ to determine the trace impedance.

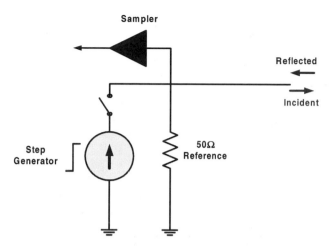

FIGURE 11.7 Simplified example of a current-source TDR output stage, including sampling input to oscilloscope.

The simplified generic example of a TDR driver output and receiver is illustrated in Figure 11.7. This contains a step generator, sampler, and reference 50-Ω termination. This illustration exemplifies a current source output stage. It should be noted that the output could be a voltage source as well, with a series Thévenin resistance. The sampler represents the receiver and is used to capture the reflected response as an input to the oscilloscope.

So now that the basic theory and operation of the TDR has been covered in the following section we go through some representative examples of extracting the basic impedance and velocity characteristics. In most practical situations the instrument TDR output is connected to a 50-Ω cable. Launching a pulse into a known cable of 50 Ω yields an incident wave with an amplitude of $\frac{1}{2}$ V, due to the resistive divider effect between the cable and the 50-Ω source resistance. The waveform reflected from the load is delayed by two electrical lengths and is superimposed back at the TDR output.

An example of a TDR response, including the cable and terminated trace (DUT), is illustrated in Figure 11.8. For this case the cable, which is assumed to be an ideal 50 Ω, is connected to a 60-Ω trace that is terminated with a 60-Ω resistor. The respective ideal TDR response shows the incident and reflected voltages used to calculate Z_{DUT}. This also shows the round-trip delay, T_d, for the overall system as well as for each individual section.

Another example is shown in Figure 11.9. This illustrates the effect of high- and low-impedance discontinuities along a 50-Ω trace. For the capacitive discontinuity loading case, the voltage decreases, and in the inductive discontinuity case, the voltage increases. Figure 11.10 represents lumped-element discontinuities. The lumped responses shown in Figure 11.10 illustrate the inductive and capacitive response for small discontinuities such as vias and PCB trace bends.

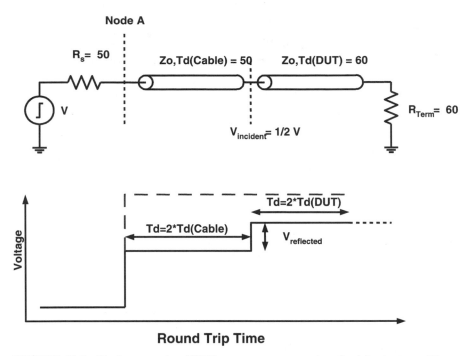

FIGURE 11.8 Basic example of TDR response as seen at node A for test conditions. The response illustrates the voltage and time relationship as the pulse propagates to termination.

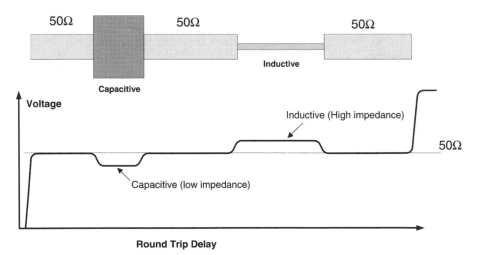

FIGURE 11.9 Simplified response illustrating the TDR response effect to capacitive and inductive loading along a 50-Ω line.

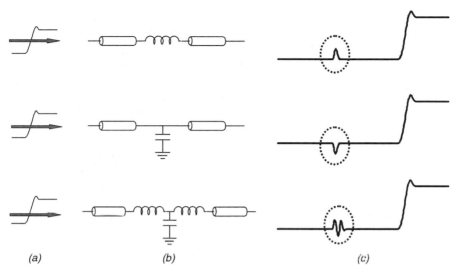

FIGURE 11.10 Typical lumped element response characteristics: (*a*) TDR input; (*b*) lumped element; and (*c*) response.

11.2.2. Measurement Factors

To properly characterize and interpret TDR response characteristics as related to the DUT (device under test), the real-life factors that affect TDR resolution in the laboratory must be understood. Insufficient resolution will produce misleading results that can miss details of impedance and velocity profiles. This is common when items to be measured are very closely spaced or when the impedance discontinuities are electrically very short. The major resolution factors for TDR are rise time, settling time, foot, and preshoot, as illustrated in Figure 11.11.

Rise Time. The most prominent TDR characteristic that affects resolution is the pulse rise time. As discussed previously, the response is based on the reflection. The general rule of thumb is that the reflection from two narrowly spaced discontinuities will be indistinguishable if they are separated by less than half the rise time. Subsequently, the TDR can only accurately resolve structures that are electrically long compared to the rise time.

$$\text{TDR}_{\text{resolution}} \geq \frac{T_{\text{rise}}}{2} \tag{11.5}$$

System rise (T_{rise}) time is characterized by the fall or rise time of the reflected edge from an ideal short or open at the probe tip.

    Faster rise times usually provide more versatility for analyzing small electrical parasitics but are also harder to interpret correctly. Most systems designs

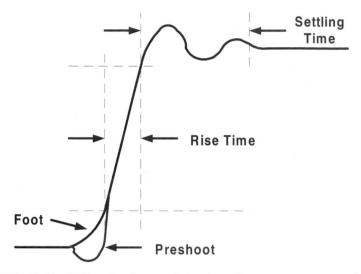

FIGURE 11.11   TDR pulse characteristics that affect measurement resolution.

may operate at a much slower edge rate than the TDR output. Subsequently, the peaks and dips of the impedance discontinues measured with a TDR will not necessarily represent those seen in the system because the slower edge rate will tend to mask small variations. For example, consider two microstrip PCB traces from two different boards that are connected together with a typical inductive connector. The connector is soldered to the main board, while the second board fits into the connector similar to a PCI or ISA card. The ISA or PCI board has pads called *edge fingers* that are exposed and fit in the connector to make electrical contact. To characterize the connector between the two boards, the overall lumped electrical model must include the connector, via, and the edge finger pad capacitance as illustrated in Figure 11.12. Figure 11.13 shows the respective TDR response of the connector for different edge rates. Notice how the effect of the inductive impedance discontinuity decreases with slow edge rates. The effect of edge rates on reflections is explained in Section 2.4.3.

Incident Characteristics. The inherent TDR response characteristics, such as settling time, foot, and preshoot (see Figure 11.11), are also important to consider with respect to TDR resolution. The incident response characteristics are essentially the difference between the ideal and actual step response. The foot and preshoot refer to the section of the response prior to the vertical edge, while settling time is the time for the response to settle.

These characteristics are inherent in the incident wave and will affect the resolution of the measurement. A simple example of this can be illustrated with the settling time. If the settling time is long compared to the electrical delay of the structure being measured, it may induce false readings because the

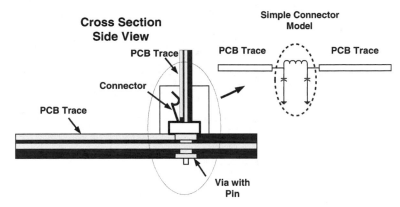

FIGURE 11.12 Cross-section view of two different PCB boards connected by a typical connector along with the simple model.

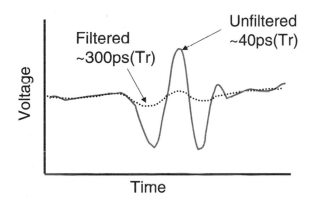

FIGURE 11.13 The TDR response for two different edge rates illustrates the resolution difference on the connector lumped elements.

impedance measurement might be taken at a peak or dip in the response. This would look like an impedance variation when in reality it is simple inherent ringing in the response. Possible methods to minimize these effects are covered in the following section.

11.3. TDR ACCURACY

TDR accuracy depends on two main components: equipment-related issues and the test system. The equipment issues include items such as step generator and sampler imperfections that are controlled by the TDR manufacturer. The test system refers to the measurement setup that incorporates cabling, probes, and type of DUT to be tested. Launch parasitics, cable losses, mul-

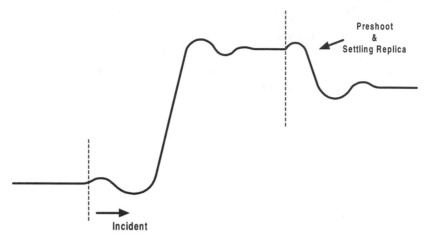

FIGURE 11.14   Example of how the incident pulse characteristics will be replicated.

tiple reflections, and interface transmission loss are all factors that can affect TDR accuracy.

For the purposes of this section, only the test system is covered, as the end user has minor control over the equipment's accuracy factors. In most cases the test system is the most critical factor in taking accurate TDR measurements. As discussed earlier, the TDR extracts impedance information based on the voltage response in time. Under ideal conditions, the response is flat during the time of interest. In the real lab environment, the response may not be so flat and could therefore affect the accuracy of the measurement. These nonideal effects along the TDR response are referred to as *aberrations*.

In some cases, aberrations can be a significant percentage of the total response. Aberrations due to the preshoot and foot of the response can be predictable. Predictable responses can use processing techniques with software to account for these effects. Unpredictable responses due to ringing and multiple reflections are not accountable. An example where the response is predictable is illustrated in Figure 11.14. TDR pulse response exhibits the foot, preshoot, and settling effects. These characteristics are also replicated in the reflected portion of the response. Therefore, since the incident response characteristics will be replicated in the reflected response, the possibility of utilizing software or digital signal processing within the TDR could be used to help eliminate some inaccuracies due to the predictable incident characteristics.

11.3.1. Launch Parasitics

One cause of unpredictable response characteristics is launch parasitics. Launch parasitics cause aberrations in the step response that can be unpredictable and therefore may not be repeatable. The goal of launching a fast edge

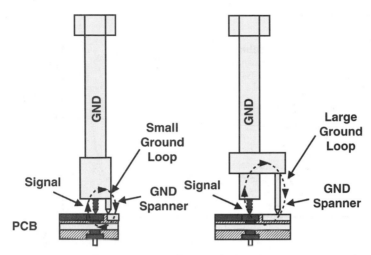

FIGURE 11.15 Examples of different hand-held type probes with varying ground loops.

while maintaining signal integrity is largely affected by reliable physical connection to the DUT, probe ground-loop inductance, and structure crosstalk. Ground-loop inductance effects due to the probe type used as an interface between the DUT and TDR are critical. Large impedance discontinuities between the TDR instrument and DUT will cause significant aberrations in the TDR response. Increasing the ground loop of the probe increases the effective inductive discontinuity between the instrument and the DUT. The larger the discontinuity, the larger aberrations that will be seen in the response, which leads to decreased accuracy.

Figure 11.15 illustrates two different examples of common hand-held probes with varying ground loops that are dependent on the probe grounding mechanism, called the *ground spanner*. It is can be seen that the probe with the long spanner will have a larger ground loop than the probe type with the small ground spanner. The larger ground loop probe will exhibit a higher-impedance discontinuity, due to the ground return path, which will make the probe response look inductive. The smaller ground-loop probe will have an inductive effect that is much smaller in magnitude than the large ground-loop probe.

Figure 11.16 shows the various response characteristics for different types of probes with varying ground-loop effects. It shows the importance of being able to control the discontinuity by minimizing the ground loop. Improved TDR response characteristics will be achieved with probe types that minimize the discontinuity between the instrument and the DUT. It should also be noted that the large ground loops tend to degrade the TDR edge. This will have a negative effect on the accuracy of the impedance and delay measurements. Minimizing the probe ground loop will generally improve settling time

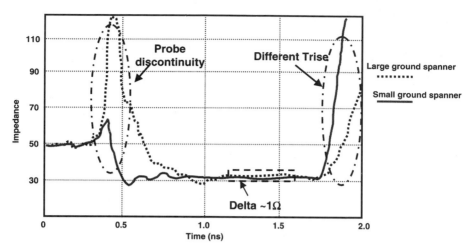

FIGURE 11.16 This illustrates a typical TDR response difference between two different handheld probes with different ground loops.

and minimize edge rate degradation, thereby increasing the resolution of the measurement.

11.3.2. Probe Types

The most commonly used probes in use today are the hand-held, surface-mounted adapter (SMA) and the microprobe. The primary factors that need to be considered when choosing the probe are accuracy and the amount of work involved to complete the measurements. The difference between probes will play a critical role in accuracy and repeatability.

The probe's purpose is to provide a connection between the measurement equipment and the DUT. Anytime the probe is not exactly 50 Ω, an impedance discontinuity will occur between the TDR output and the load under test. This discontinuity will cause a reflection at the probe and decrease the accuracy of the measurements. Minimizing this reflection requires that the probe exhibit a continuous impedance of 50 Ω all the way to the probe tip (because it matches the cable and the output impedance of the equipment). The impedance mismatch between the cable and the probe is one of the primary factors that affect measurement accuracy. The smaller the discontinuity, the greater the accuracy [McCall, 1999].

The most commonly used probe type is the hand-held probe. This probe is useful for quick checks but is not adequate for the most accurate or repeatable measurements. A common problem with hand-held probing is measurement error due to the ground connection mechanism of the probe (the ground spanner). Selecting the proper ground spanner is critical to minimizing the discontinuity, due to the probe ground-loop inductance. Minimizing the ground-loop inductance reduces the initial spike (high impedance) as the pulse launches into the load under test. The benefits of minimizing the ground loop are improved settling time, response smoothness, and decreased probe loss.

SMA connectors soldered to the board will provide good repeatability because they are soldered onto the board. The problem is that they tend to exhibit rather high capacitive loading, which reduces the effective edge rate of the TDR pulse into the device under test. SMA connectors are not good for measurement conditions that need to utilize maximum resolution of TDR instruments [McCall, 1999].

Controlled impedance microprobes are the most accurate probe type that can be used. Microprobes exhibit very small parasitics, very small ground loops, and are usually 50 Ω all the way to the probe tip (although high-impedance microprobes can also be purchased). Microprobes are generally used for characterizing radio-frequency and microwave circuits. These are predominately used in the industry with the vector network analyzer (VNA), which is explained later in this chapter. However, in the digital world they can be used for the most accurate TDR measurements. The major disadvantage of these probes is that they require a probe station, special probe holders, and a microscope [McCall, 1999].

11.3.3. Reflections

Multiple reflections in the test system may or may not be predictable. Remember that the TDR response is a reflection profile, not an impedance profile. Also remember that reflection magnitude and the resolution will be a factor of rise-time degradation. If the resolution is lost, the predictability of reflections will also be lost. The most accurate line measurements are usually completed on lines that have no discontinuity, such as impedance coupons. A good rule of thumb is to minimize reflections prior to the DUT and any impedance discontinuities between the TDR and DUT. This requires the engineer to use a controlled-impedance, low-loss cable as the interface between the TDR and the probe. Furthermore, a low-loop-inductance, controlled-impedance probe is required. To understand the effect of multiple reflections due to multiple impedance discontinuities, one can apply a lattice diagram. This is shown in Figure 11.17, illustrating the superposition of reflections for multiple impedance discontinuities.

11.3.4. Interface Transmission Loss

Interface transmission loss is a big factor, especially in test system environments that are significantly different than the instrument 50-Ω reference. When the test system impedance is significantly different from the DUT, only a portion of the initial energy-launched into the interface is transmitted. Therefore, a reflection with a smaller amplitude is passed back through the interface to

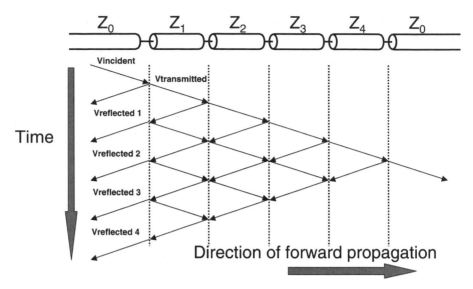

FIGURE 11.17   Lattice diagram of multiple discontinuities, resulting in the superposition of reflections.

the source. This will make it difficult to measure impedance's significantly different than 50 Ω.

11.3.5. Cable Loss

Most test systems will have a cable interface between the instrument and the probe. Cable loss is another factor that can reduce accuracy, by causing apparent shifts in characteristic impedance. This is due primarily to skin effect losses, which create a pseudo-dc shift in the line impedance. Amplitude degradation will occur as the pulse propagates down a lossy line, resulting in an error between the actual incident amplitude input versus the output of a cable. This inaccuracy can either be eliminated by the use of expensive low-loss cables or compensated following the proper measurement techniques that will be covered later in the TDR accuracy improvement sections.

11.3.6. Amplitude Offset Error

Amplitude offset error is when the incident voltage amplitude varies versus the reference voltage. A common effect that is noticed on some TDR measurements is a reading variation, depending on whether or not the DUT is terminated. If the impedance of a cable, for example, is measured, the reading may change, depending on whether the cable is left open or is terminated. It is possible that the reading could exhibit a 1- or 2-Ω delta between the open-ended and terminated case. This is a result of providing a dc path to ground that simply loads the step generator driver, causing a reduction in the incident step amplitude. To eliminate the source of error for general impedance measurements, it is always recommended to leave the DUT open ended.

11.4. IMPEDANCE MEASUREMENT

The following procedures will aid the engineer to account for the major sources of TDR inaccuracy, resulting in very accurate impedance measurements [IPC-2141, 1996]. These measurements utilize a reference standard airline or golden standard, of which the impedance and delay has been carefully characterized. It is possible to purchase airline standards that are NIST (National Institute of Standards Test) certified to achieve the greatest accuracy to less than $0.1\ \Omega$. To achieve the most accurate results, the impedance of the reference standard should be very close to the test system characteristic impedance to minimize interface transmission losses. For example, if the test system environment design was nominal $30\ \Omega$, a reference standard with the same impedance should be chosen.

Figure 11.18 shows the basic test setup. This includes the cable, probe, DUT, and airline standard. Both the airline standard and the DUT must be open ended. The probe type, for the purpose of this example, should have a controlled impedance that matches the cable and should exhibit very small parasitics.

11.4.1. Accurate Characterization of Impedance

As described above, there are several sources of error that will decrease the accuracy of an impedance measurement. Described below are two different ways of measuring the impedance accurately and negating some of these errors.

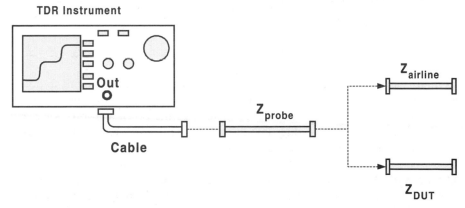

FIGURE 11.18 Basic test setup for accurate impedance measurements using an airline reference standard.

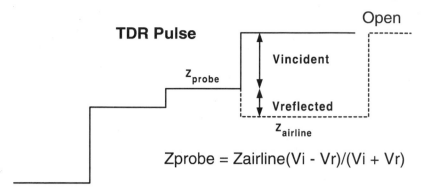

FIGURE 11.19 Measurement response for determining Z_{probe}.

Back Calculation Using the Standard Reference. The first step is to determine the probe impedance. This is determined by measuring the step size of the incident response at the probe tip. This is done with the open-ended probe attached to the end of the cable as shown in Figure 11.18. Next, the airline is connected to the cable, and the reflected step response is measured as shown in Figure 11.19. The probe impedance can be calculated by using the equation

$$Z_{\text{probe}} = Z_{\text{airline}} \frac{V_I - V_R}{V_I + V_R} \qquad (11.6)$$

where V_I is the incident step response, V_R the reflected step response, and Z_{airline} the known impedance of the standard airline.

Now that the probe impedance is determined, we can calculate the actual DUT impedance by following a similar procedure. Measure the incident step size with the probe open and then measure the reflected step response while probing the DUT. The DUT characteristic impedance is extracted using the equation

$$Z_{\text{DUT}} = Z_{\text{probe}} \frac{V_I + V_R}{V_I - V_R} \qquad (11.7)$$

In many cases the discontinuity of the probe being used is too small to determine the impedance. Completing the same measurement as discussed previously using the cable as the reference instead of the probe can provide good results, on the order of less than 1-Ω difference. This is especially true in cases where the probe impedance effects are quite small.

Offset Method. Another technique of measuring impedance is the offset method. The offset method simply uses a fixed value that accounts for the difference between the standard and the actual readings. A standard characterized impedance reference (such as the airline) is measured and the difference between the measurement and the known impedance is the offset. When measuring the DUT, the offset is applied to the measurement result to account for errors. This technique is especially useful for inexperienced users. This assumes a linear relationship within the given specification window which is not necessarily correct. Although inferior to the back-calculation method, it does produce relatively accurate results with less time [McCall, 1999].

Impedance Coupon. Depending on the topology of the bus design, it may or may not be possible to measure the actual signal traces to determine the bus characteristic impedance. This will depend on the length of the traces and bus impedance. The larger the difference in Z_o between the DUT and TDR, the longer the settling time to obtain a flat response. Furthermore, it may not be possible to probe the signal traces on the final board design without using a probe with a large inductive ground loop. To circumvent this problem, an impedance coupon can be used. In PCB manufacturing, impedance coupons are used to dial in the process to obtain the target impedance. To characterize the characteristic impedance, it is important that the coupon geometry replicates the bus design. It is also critical that the coupon is designed properly for obtaining accurate measurements.

11.4.2. Measurement Region in TDR Impedance Profile

Measurement methodology can play a big role in correlation and accuracy. For example, the TDR response in Figure 11.20 illustrates measurement of a 6-in. low-impedance coupon using a common hand-held type of probe. Notice the initial inductive spike and the amount of time it takes the response to settle. The purpose of this illustration is to show the possible variations in the results due to cursor positioning along the response. Measurements that are taken during the settling time will contain more variability and will therefore be prone to incorrect readings. To minimize the variability of measurement location along the response, a good rule of thumb is to measure the mean of the response along the flattest region, as highlighted in Figure 11.20. This works well provided that the structure is long enough to allow for sufficient settling time.

The second item to note is that trying to measure a short trace would lead to incorrect impedance values because there is insufficient time for the response to settle. In the case of Figure 11.20, the measurement of the 6-in. trace takes several hundreds picoseconds to settle. If an identical 1-in. trace, for example, was characterized, the ringing from the inductive probe would not have time to settle before the reflection from the open at the end of the trace returns to the sample head. Subsequently, there would be no flat areas of the response to measure, and it would be very difficult to achieve accurate results. This is a common error made in today's industry. As mentioned previously, proper coupon design is also critical for accurate impedance measurements.

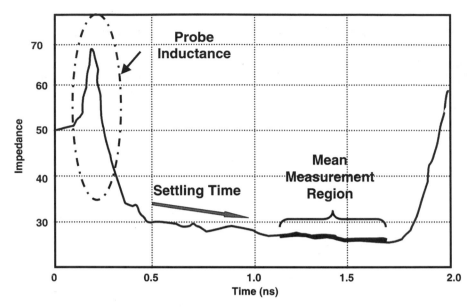

FIGURE 11.20 Measurement of low impedance 6 in. microstrip coupon with an inductive probe.

Test coupons should be designed to replicate the bus design and therefore should take into account trace geometry, copper density, and location on the PCB.

RULES OF THUMB: Obtaining Accurate Impedance Measurements with a TDR

- Complete measurements with the DUT open circuited. This eliminates step amplitude variation due to dc loading.
- Obtain measured data in the flattest region of the TDR response. This should be far away from the incident step, where aberrations will be most pronounced.
- Try to minimize probe discontinuities and cable loss. Use of small, controlled impedance microprobes along with low-loss cables will minimize these sources of error.
- Utilize test structures that are long enough to allow the TDR signal to settle.
- When designing an impedance coupon, try to replicate the topologies that will be used in the design.

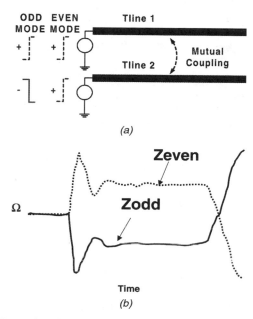

FIGURE 11.21 Example of a TDR response for a differential impedance measurement of two traces left open. (*a*) differential measurement setup (requires two simultaneous sources that are phase aligned in time); (*b*) differential TDR response.

11.5. ODD- AND EVEN-MODE IMPEDANCE

Differential impedance measurements are required to extract the odd-mode impedance between two signal lines that are coupled. To complete differential measurements, two TDR sources need to be injected across a pair of coupled signal lines simultaneously. The TDR output edges must be phase aligned so that the output edges have zero time delay (skew) between them. After aligning the edges, reverse the polarity on one source to measure the odd-mode impedance or leave the polarity the same to extract the even-mode impedance. An example of this is shown in Figure 11.21.

11.6. CROSSTALK NOISE

Another important parameter that must be characterized in high-speed design is the voltage induced on adjacent lines due to crosstalk. Similar to differential impedance measurements, crosstalk is measured utilizing two or more channels. A common method used to evaluate crosstalk is to measure the noise induced on a passive line due to the neighbor or neighbors. For this case, two coupled lines will be examined.

Crosstalk can be measured as illustrated in Figure 11.22 by injecting the TDR pulse into a line and measuring the noise induced on the victim line. The

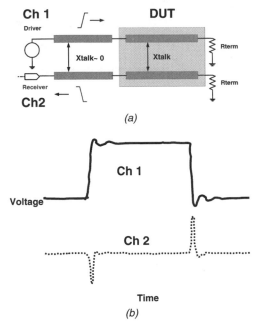

FIGURE 11.22   Example of (*a*) a basic crosstalk measurement setup along with (*b*) the response.

chart shown in Figure 3.4 will relate the magnitude and shape of the crosstalk pulses for a two-line system to the mutual parasitics, input step voltage, edge rate, and victim termination. Completing simulations of the same net can be done to compare modeling accuracy against lab data. Filtering functions can be implemented to examine noise for slower rise times to emulate system bus speeds. Filtering functions are predominantly digital filters that postprocess the data received within the TDR instrument. The use of passive filters inserted inline are not recommended.

11.7. PROPAGATION VELOCITY

Measurement of propagation delay and velocity is more difficult than measuring impedance. For velocity, the structure delay is determined by measuring the difference in time that it takes the pulse to propagate through the structure. Measurement points for propagation delays are difficult to choose and accuracy is extremely dependent on the probing technique. The most accurate delay measurements require advanced probing techniques that utilize controlled-impedance microprobes with the TDR in time-domain transmission (TDT) mode. This improved accuracy comes at a cost of equipment and time. Selecting the method for measuring the velocity depends on the accuracy desired and the amount of time allotted for measurement. It should be noted

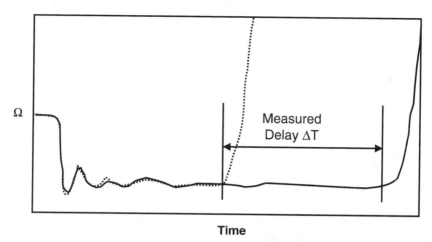

Time

FIGURE 11.23 Example of a TDR propagation delay measurement for two identical impedance coupons that differ only in length (Adapted from McCall [1999].)

that velocity measurements are to be used to extract the effective dielectric constant as shown with equations (2.3) and (2.4).

11.7.1. Length Difference Method

The simplest method for measuring propagation delay is using TDR mode to measure delay between two identical test structures of different length (TDR mode assumes that the user is looking only into the near end of the DUT). Velocity is determined by subtracting the reflected delay differences between the two structures. As shown in Figure 11.23, the ΔT value will be twice the actual delay difference due to reflection delay down and back. The velocity per unit length is calculated as

$$\text{velocity} = \frac{L_1 - L_2}{\frac{1}{2}\Delta T} \tag{11.8}$$

The purpose of using two test structures is to cancel out errors in the determination of the point where the waveform begins to rise due to the open circuit at the far end of the DUT. Accuracy will depend on probes and structure. To make this measurement, maximize the TDR time base and position the cursors at the point where the reflected pulse begins to rise, as illustrated in Figure 11.23.

11.7.2. Y-Intercept Method

If space permits, the approach described above can be improved by using three test structures of different length. To calculate the velocity, plot length versus delay (length on the x-axis). Draw a line that that best fits the data (if the measurements are taken consistently, the three points should lie on a straight line). The inverse of the slope will be the propagation velocity. If the measurement is perfect, the line will intercept the y-axis at zero. Otherwise, the actual y-axis intercept gives the measurement error.

11.7.3. TDT Method

Improved accuracy of propagation delay measurements can be completed with the TDR used in TDT mode (TDT means that the signal is also observed at the far end of the line). In this method, probes are placed at each end of the test structure. A pulse is injected into one end and captured at the other end. This approach has less edge-rate degradation than the simpler methods already mentioned, which results in improved accuracy. The measurement is completed by injecting the pulse on one end of the test coupon with a 50-Ω probe and capturing the signal at both the launch point and the open end with a low-capacitance high-bandwidth high-impedance probe ($10\times$ or $20\times$), as illustrated in Figure 11.24. The advantage of this technique over the TDR is that the captured signal has propagated only once down the coupon when it is captured, which will allow less time for the edge rate to be degraded by losses or reflections.

The measurements are completed by connecting the 50-Ω probe to a sampling head with the TDR/TDT mode "ON." The high-impedance probes should be connected to separate channels and with the head function's TDR/TDT setting mode to "OFF." Measure the delay between the transmitted and received signals (using the high-impedance probes at the driver and receiver). This is the propagation delay of the trace [McCall, 1999].

All methods of measuring propagation velocity are highly subjective, due to where the measurement point along the response is taken. The ideal measurement point is that at which the response just begins to rise. This would eliminate the error due to different edge-rate degradation if measurement

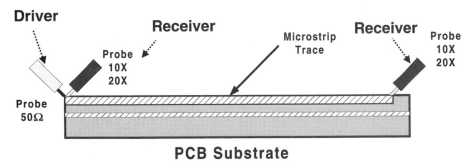

FIGURE 11.24 Example of a basic TDT measurement setup for evaluating propagation velocity of a microstrip trace.

were taken at, for example, the 50% point. However, in most circumstances the incident preshoot and foot characteristics can make it quite difficult to determine exactly when the response is rising [McCall, 1999].

11.8. VECTOR NETWORK ANALYZER

One of the prominent laboratory instruments used for radio-frequency and microwave design purposes is the vector network analyzer (VNA). A VNA is ideally suited for measuring the response of a DUT (device under test) as a function of frequency. The primary output of a VNA is reflected and transmitted power ratios or the square roots thereof. Through mathematical manipulation of the power ratios, a large amount of valuable data can be extracted. A few examples that can be measured with a VNA are skin effect losses, dielectric constant variations, characteristic impedance, capacitive and inductive variations, and coupling coefficients. Furthermore, since VNAs are typically designed with the microwave engineer in mind, they can make reliable measurements to extremely high frequencies. The primary disadvantage is that they provide data in the frequency domain. Subsequently, it is necessary to translate the extracted data into a useful format that can be used in the time domain.

In the past, most digital designs had no need for extracted electrical characteristics in the hundreds of megahertz range, simply because their operating frequencies were low compared to today's systems. Modern bus designs, however, are becoming so fast that such measurements are a necessity. As bus speeds pass the 500-MHz region, rise and fall times are required to be as fast as 100 ps. Subsequently, the frequency content of the digital waveforms can easily exceed 3 GHz. As bus frequencies increase, the resolution required for measuring capacitance, inductance, and resistance increases. To the design engineer, proper characterization to a fraction of a picofarad, for example, might be necessary to characterize the timings on a bus accurately. If used properly, the VNA is the most accurate tool available to extract the electrical parameters necessary to simulate high-speed designs.

The primary focus of this discussion is on VNA operation to extract relevant electrical parameters prominent in digital high-speed bus designs. This includes propagation delay, crosstalk, frequency-dependent resistance, capacitance, and inductance parameters. Basic procedures used to measure items such as connectors, sockets, vias, and PCB traces are covered. It should be noted that these techniques may or may not apply to the reader. However, for the novice VNA user, these procedures should help provide some helpful concepts for utilizing the VNA to yield accurate results.

Because the focus of this chapter is on validating and developing models for high-speed digital design, the measurement techniques discussed are limited to the common methods most useful for the characterization and development of models. The discussion includes measurement techniques, sources of error, and instrument calibration.

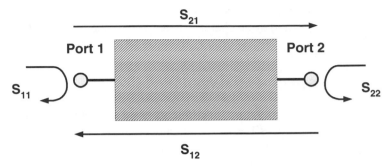

FIGURE 11.25   Two-port junction. (Adapted from Pozar [1990].)

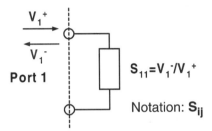

FIGURE 11.26 S-parameter one-port definition and general notation.

11.8.1. Introduction to *S* Parameters

The primary output of the VNA is the *scattering matrix*, also know as *S parameters*. The scattering matrix provides a means of describing a network of N ports in terms of the incident and reflected signals seen at each port. As frequencies approach the microwave range, the ability to extract voltage and currents in a direct manner becomes very difficult. Under these circumstances, the only directly measurable quantities are the voltages (or power) in terms of incident, reflected, and transmitted waves, which are directly related to the reflection and transmission coefficients at the measurement ports. Therefore, the scattering matrix is defined in terms of the reflected signals from each port of a device and the transmitted signals from one port to another, as shown in Figure 11.25.

The general nomenclature is this. S_{ij} means that the power is measured at port i and injected at port j. For example, S_{11} is defined as the square root of the ratio between the power reflected at port 1 over the power injected into port 1. This can usually be simplified by using the voltage ratios, as shown in Figure 11.26. Subsequently, S_{11} is defined as the reflection coefficient of the ith port if $i = j$ and all other ports are impedance matched.

Equation (11.9) is a simplified definition of S_{11} when the conditions stated above are met.

$$S_{11} = \sqrt{\frac{P_{\text{reflected},1}}{P_{\text{incident},1}}} = \Gamma = \frac{V_{\text{reflected},1}}{V_{\text{incident},1}}. \tag{11.9}$$

The transmission coefficients, such as S_{21}, are defined as the output voltage at port 2 over the input voltage at port 1 and can be used to measure items such as propagation delay and attenuation:

$$S_{21} = \sqrt{\frac{P_{\text{transmitted},2}}{P_{\text{incident},1}}} = \frac{V_{\text{port2}}}{V_{\text{port1}}} \tag{11.10}$$

For symmetrical systems, such as a PCB trace, $S_{22} = S_{11}$ and $S_{12} = S_{21}$. If the network is lossless, the S-parameters must satisfy the equation

$$|S_{11}|^2 + |S_{12}|^2 = 1 \tag{11.11}$$

This is a useful relationship for low-loss systems and at frequencies when skin effect resistance is small. If these characteristics hold true for the DUT, the networks can be simplified greatly to reduce the calculations needed to extract the parameters.

11.8.2. Equipment

The VNA measurement setup needed to achieve high accuracy will require the use of controlled impedance microprobes along with their respective calibration substrate. This will therefore require a stable probe station along with optics (a microscope) to achieve good repeatable measurements that will minimize the possibility of expensive equipment damage.

11.8.3. One-Port Measurements (Z_o, L, C)

One-port VNA measurements are useful for extracting inductance, capacitance, and impedance as a function of frequency. This is especially useful for the characterization of connectors, vias, and transmission lines. The S-parameter one-port measurement is a measure of S_{11} or S_{22}.

To understand how inductance and capacitance are measured at low frequencies, consider a short section of a transmission line. A low frequency, for these applications, is when the wavelength is much larger than the length of the structure. The relationship between wavelength and frequency is $\lambda = c/(F\sqrt{\varepsilon_r})$, where F is the frequency, c the speed of light in free space, and ε_r the effective dielectric constant.

By solving the circuit equations we can derive the voltage and currents for a short length of open and shorted transmission line:

$$V = (R + j\omega L)I \quad \text{(short circuit)} \tag{11.12}$$

$$I = (G + j\omega C)V \quad \text{(open circuit)} \tag{11.13}$$

If we consider only the first-order effects by neglecting R and G, the two equations simplify so that a current flowing through the line will produce a voltage that is proportional to the inductance if the line is shorted to ground. Furthermore, the voltage induced between the line and ground will produce a current that is proportional to the capacitance if the line is left open ended. To understand when these approximations apply, let us first consider the case when the structure is shorted to ground. As long as the impedance of the structure capacitance is large compared to the impedance of the structure inductance at the measurement frequency, there will be minimal current flowing through the capacitor. Subsequently, the voltage drop across the structure will be primarily a function of the inductance when the structure is shorted to ground. Now, let us consider the case when the structure is left open ended. As long as the structure is small compared to the wavelength at the measurement frequency, it will look like a lumped element capacitor. Subsequently, the structure will charge up and the current will flow to ground through the structure capacitance. The structure inductance can be ignored as long as the wavelength is sufficiently long so that no significant transient voltage differentials can exist across the structure. As a result, under the conditions stated above, the current will depend primarily on the structure capacitance.

By opening and shorting the far end of a DUT, one-port VNA measurements (i.e., S_{11} or S_{22}) are made to extract the inductance and capacitance of a short structure, such as a connector pin. If the VNA is set to display a Smith chart, as shown in Figure 11.27, the inductance and capacitance values can be determined most easily. To determine the capacitance, the DUT should be open ended. To determine the inductance, the far end of the DUT should be shorted to ground.

A *Smith chart* is a graphical aid that is very useful for visualizing frequency-domain problems. This display format is used to visualize normalized impedance (both real and imaginary parts) in terms of the reflection coefficient as a function of frequency. The normalized impedance is equivalent to the impedance of the DUT divided by the VNA system impedance, which is set during calibration. The VNA system impedance is normally 50 Ω; however, it can be set to alternative values when necessary to provide more accurate results when the target DUT impedance is significantly different. The center point (1), shown in Figure 11.27, corresponds to the VNA output impedance $(1 + j0)Z_{\text{VNA}}$. The horizontal line through the center for the Smith chart corresponds to the real impedance axis, which goes from zero ohms (short) on the left-hand side through to infinite impedance (open) on the right-hand side. Furthermore, the outside perimeter equates to purely reactive impedance values

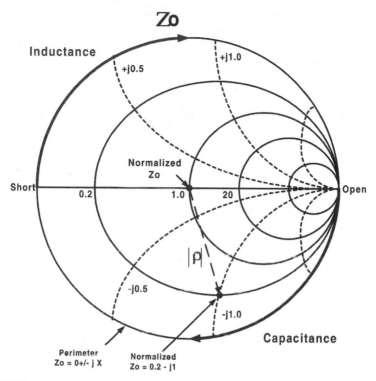

FIGURE 11.27   Representative Smith chart highlighting the basic parameters.

such as purely capacitive or inductive loads. Several references exist explaining the Smith chart in greater detail. The reader is encouraged to learn more about the chart since it is a very useful tool. For the purposes here, only points on the perimeter of the chart will be considered.

Measuring Self-Inductance and Capacitance of a Short Structure. To show briefly how the reactive components are extracted, let's consider a small trace or stub terminated with a short to ground (assume that the losses are negligible). Since the termination is a short, the reflection coefficient is −1 and the impedance seen will be reactive (because it will lie on the outer perimeter of the Smith chart). For low frequencies, this structure will look inductive as long as the trace is electrically shorter than $\lambda/4$ of the frequency of interest. When the frequency is very low, the DUT will look almost like a pure short to ground. As the frequency increases, the impedance will increase ($j\omega L$), and the value on the Smith chart will move clockwise toward the open point, as depicted in Figure 11.27.

The inductance can be calculated by observing the imaginary portion of the impedance on the Smith chart (the real part should be very small) and solving equation (11.14) for the inductance, where ω is the angular frequency ($2\pi F$).

The reader should pay close attention to whether or not the value read off the Smith chart is normalized. The VNA output may or may not be set to display normalized data. The equation

$$Z_{\text{inductance}} = j\omega L \qquad (11.14)$$

assumes that the VNA reading is not normalized to the VNA system impedance. If the same measurement is taken with the DUT open ended, the reflection coefficient will be 1 and the trace will look capacitive. At low frequencies, the Smith chart cursor will be very close to the open point in Figure 11.27 and will move clockwise toward the short as frequency increases.

The capacitance can be calculated by observing the imaginary portion of the impedance on the Smith chart (the real part should be very small) and solving the equation

$$Z_{\text{capacitance}} = \frac{-j}{\omega C} \qquad (11.15)$$

for the capacitance. Again, pay attention to whether or not the data is normalized. Equation (11.15) assumes nonnormalized data. If the DUT exhibits significant losses, or the frequency gets high enough so that the assumptions above are violated, the trace on the Smith chart will begin to spiral inward because the impedance is no longer purely imaginary.

For more accurate results, some measurements may require two identical structures that vary only in length. For example, if there are vias on a test board that are required to interface the probe to the DUT, they may interfere with the measurement. If the parasitics of the vias, or probe pads, are significant compared to the DUT, this will cause a measurement error. If two short structures of different lengths are measured, the difference in the measured parasitics can be divided by the difference in the length to yield the inductance or capacitance per length. This will produce more accurate results because the parasitics of the probe pads and vias will be subtracted out.

Although these measurement techniques are valid only for low frequencies, the results can generally be applied to much higher frequency simulations because the capacitive and inductive characteristics of a structure (such as a socket or connector) will typically not exhibit significant variations with frequency.

Socket and Connector Characterization. The primary metric used for characterizing a connector or a socket is the effective inductance and capacitance. As described in Chapter 5, the equivalent parasitics depend on the physical characteristics of the pin, but also on the pin assignment. This is illustrated in Figures 5.5 and 5.6. Subsequently, when measuring the inductance and capacitance of a connector or socket, it is imperative that the measurements be performed in such a manner so that the results reflect, or can be manipulated, so that the effective parasitics seen in the system can be obtained. To illustrate this, refer to Figure 5.6*a* and *d*. Significantly more coupling to the adjacent data pins will be experienced in Figure 5.6*a* than in Figure 5.6*d*, whose pinout isolates the signals from each other. Although the self-inductance of the pins in this example may be identical, the effective inductance as seen in the system will be quite different.

To measure or validate the DUT, specific test boards are usually required that have open and short structures so that the capacitance and inductance can be measured. In instances when electrical values of the test structure used to interface the probe to the DUT are significant (e.g., a trace on the test board leading from the probe to the socket/connector), two different measurements are required. First, the equivalent parasitics of the trace and probe pad leading to the DUT must be determined. Next, the total structure, including the interface traces and the DUT, must be measured. By subtracting the trace parasitics from the total parasitics, the DUT self-electrical values can be determined.

An example of a connector test board is illustrated in Figure 11.28. In this instance there are structures that are used to interface the microprobe to the connector. Since the goal is to extract the electrical parameters of the connector, not the interface structures on the test board, open and short measurements will need to be completed first to extract the *L* and *C* values of the test board without the connector. Next, the measurements are taken to obtain the total *L* and *C* values of the connector plus the test board. The two results are subtracted, resulting in the *L* and *C* values of the connector. This typically requires two separate test structures, one to extract the board parasitics and one to extract the total board and DUT parasitics. These structures should be placed in close proximity to each other to minimize differences across the board due to manufacturing variations.

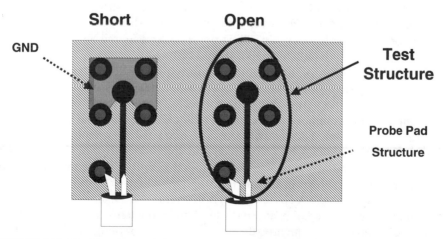

FIGURE 11.28 Simple example of a socket test board used to measure the parasitics.

One-Port impedance Measurement. Impedance measurements can be calculated based on inductance and capacitance data obtained per unit length using the self-value procedures described. An alternative method is to use the VNA in TDR mode. Most VNAs have time-domain capability using the inverse FFT (fast Fourier transform) to translate the frequency-domain data to a time-domain response. When this is done, the time-domain response will be as discussed earlier in the chapter for TDR. Therefore, measurements should be extracted in a similar manner. To achieve the highest accuracy, the VNA output impedance should be calibrated to the nominal design impedance. Another nice feature of using the VNA in the time domain is the ability to set the system bandwidth and thereby tune the edge rate to match the design.

Transmission Line (R, L, G, C). Transmission line parameters can be derived from both open and short S_{11} measurements. General common method for model derivation or validation is to utilize modeling software to extract the parasitics from the measured data. The measured data, in terms of magnitude and phase versus frequency, can be input into the modeling software to curve fit the frequency-domain response and thereby yield the *RLCG* values used to accurately model lossy, frequency-dependent models.

11.8.4. Two-Port Measurements (T_d, Attenuation, Crosstalk)

Two-port measurements such as S_{21} and S_{12} are used to measure the forward and reverse insertion loss. This is used to measure mutual terms, attenuation, and propagation delay.

Mutual Capacitance and Inductance of an Electrically Short Structure. The techniques presented here are very useful for measuring the mutual capacitance and inductance between two electrically short structures, such as connector or socket pins. Again, the structure is considered short if the electrical length (delay) is much smaller than $\lambda/4$.

The measurements of the mutual terms require two test structures that neighbor each other. Care should be taken when designing the test board so that the interface traces (the traces that connect the probe to the connector) are not coupled. One common method is to angle the traces into the structure and minimize parallelism. The mutual terms are extracted in a manner similar as to that for the self-values by using open and shorted structures. Opening and shorting the far end of the two structures together will produce the simplified equivalent circuits shown in Figure 11.29.

The two ports of the VNA are shown in Figure 11.29. For the case when the circuit is open ended, the current through the inductance will be negligible, therefore ignored. Subsequently, the main source of coupling between ports 1 and 2 is the mutual capacitance. The mutual capacitance is extracted with one

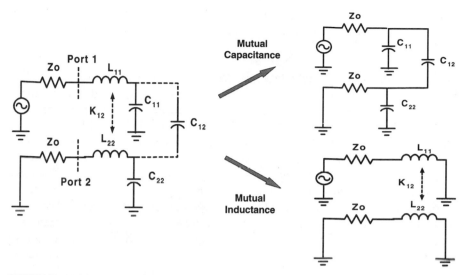

FIGURE 11.29 Basic simplification for measuring mutual inductance and capacitance.

of the following two equations:

$$S_{21} \approx \frac{2Z_o j\omega C_{12}}{1 + 2Z_o j\omega C_{12}} \tag{11.16}$$

$$\frac{1}{j\omega C_{12}} = Z_o \frac{(1 + S_{11})(1 + S_{22}) - S_{12} S_{21}}{2 S_{21}} \tag{11.17}$$

which are derived in Appendix D. Note that the value of S_{21} taken from the VNA may be in decibels and thus must be converted to linear units prior to insertion into the equations. Equation (11.16) is valid when the impedance of the shunt (self) capacitors $(1/j\omega C_{11})$ is much larger than Z_o (the VNA system impedance). This usually occurs for conductors that are perpendicular to the system reference plane. If the impedance of the shunt (self) capacitance of the structure is not high compared to Z_o, equation (11.17) may be used. Equation (11.17) is derived directly from the *ABCD* parameters discussed in Appendix D.

If each leg of the circuit is shorted to ground, the value of S_{21} will depend primarily on the mutual inductance. By shorting the two ends of the DUT to ground and simplifying the circuit as shown in Figure 11.29, S_{21} can be related to the mutual inductance as

$$S_{21} \approx \frac{2j\omega L_{12}}{2j\omega L + Z_o} \tag{11.18}$$

assuming $L_{11} = L_{22} = L$. Again, the derivation is shown in Appendix D. In equations (11.16) through (11.18), Z_o is assumed to be the system impedance of the VNA and $\omega = 2\pi F$.

Propagation Velocity and Dielectric Constant. Propagation velocity is becoming an increasingly important specification in high-speed digital designs. Modern designs require that the flight times and interconnect skews be controlled to 5 ps per inch of trace or less. Calibrated correctly, the VNA can provide a significantly more accurate characterization of the propagation delay than can be achieved with the TDR.

The VNA can measure propagation delay in two different ways. The first and less accurate method is to put the VNA in time-domain mode and perform a normal TDR/TDT measurement as described earlier in the chapter. For TDT measurements the data are taken by using a two-port measurement that compares the open S_{11} response to the S_{21} response. The more accurate method is to measure the phase of the signal using a two-port measurement. This technique simply looks at the phase difference between ports 1 and 2. The phase difference is proportional to the electrical delay of the structure. The measurement is most easily taken by displaying the data in a polar format, as shown in Figure 11.30. With the VNA in polar format, the phase delay between ports 1 and 2 can be measured, which can be used to calculate the propagation delay as a function of frequency. This is typically much more accurate than the TDR method because the VNA will output a sine wave. Losses in a system can significantly degrade the edge rate in a TDR measurement, which leads to

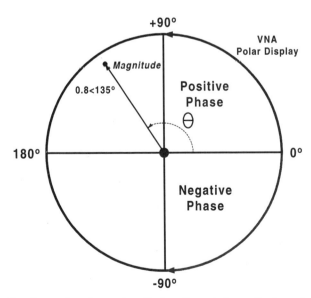

FIGURE 11.30 Example of a polar display format (magnitude and phase) versus frequency. Delay will be a function of phase and loss a function of magnitude.

incorrect delay measurements. As losses will degrade primarily the amplitude of a sinusoidal signal, they will have a minimal effect on the accuracy of the measurement.

Polar measurements are completed measuring forward S_{21} or reverse S_{12} with the VNA display in polar format. With the unit in polar format, the phase angle as a function of frequency is displayed. Adjust the marker for the frequency at which the propagation delay is desired. Delay can then be calculated using the equation

$$T_d = \Phi \frac{1}{F(360)} \tag{11.19}$$

where F is the frequency and Φ is the measured phase angle. The propagation velocity is calculated with the equation

$$v = \frac{x}{T_d} \tag{11.20}$$

where x is the length of the transmission line being measured. As shown in equation (2.4), the time delay depends on both the length and the effective dielectric constant. Therefore, if the length and propagation velocity are known, the effective dielectric constant can be extracted:

$$\varepsilon_r = \left(\frac{T_d c}{x}\right)^2 \tag{11.21}$$

where x is the line length and c is the speed of light in a vacuum.

Attenuation (R_{ac}). Attenuation measurements on a transmission line can be difficult to extract when the line impedance is different from the VNA system impedance, because reflections on the DUT will interfere with the measurement. When the system impedance is matched to the line under test, there are no reflections at the input and output port. Therefore, the value of attenuation is related simply to the magnitude of S_{21} or S_{12}. When the reflections are large, mathematical manipulation of the two-port S parameters may be required. For cases when the VNA system Z_o is close, or perfectly matched to the transmission line, the measurement becomes fairly simple. As a general rule of thumb, fairly accurate attenuation measurements can be extracted when the reflection coefficient is less than -20 or -25 dB:

$$20 \log \left| \frac{Z_{DUT} - Z_o}{Z_{DUT} + Z_o} \right| < -25 \text{ dB} \tag{11.22}$$

This is based on the typical noise floor of -60 dB set during calibration. This may or may not provide acceptable results, due to the varying degree of accuracy required by the user.

For transmission line values that are significantly different from the standard VNA impedance of 50 Ω, it may be necessary to purchase a calibration substrate with an impedance equal to the nominal impedance of the DUT. In fact, it is advisable to purchase several calibration substrates that are compatible with the specific probe types used that are in 5-Ω increments. This will allow the user to calibrate the VNA to whatever impedance is required. When the calibration substrates are not available, it is possible to make the measurements with the VNA calibrated to its nominal impedance of 50 Ω and mathematically normalize the data to the characteristic impedance of the transmission line under test. This technique is generally less accurate because small imperfections in the calibration tend to be amplified in the calculation. However, if a calibration substrate of the proper impedance is not available, it may be necessary. This technique is outlined in Appendix D.

Assuming that the VNA impedance is sufficiently close to the impedance of the DUT, the attenuation due to the conductor losses can be extracted using the equation

$$S_{21} \approx \frac{2Z_o}{R_{ac} + 2Z_o} \tag{11.23}$$

where Z_o is the impedance of the VNA and R_{ac} is the extracted resistance as a function of frequency. Equation (11.23) is derived in Appendix D. As described in Chapter 4, the skin effect resistance, up to a few gigahertz or so (assuming FR4), will dominate the attenuation. At higher frequencies, the attenuation will be a function of both the dielectric losses and the skin effect losses.

11.8.5. Calibration

As mentioned previously, the calibration procedure is critical to achieving accurate results. The purpose of this section is to provide a brief background on the VNA sources of error and discuss some calibration techniques. Calibration is done to compute the error constants used to eliminate the described sources of error. The basic VNA sources of error for one port S_{11} measurements are directivity, frequency tracking, and source matching. Illustrative examples are shown in Figure 11.31.

Directivity Error. Directivity error occurs when less than 100% of the incident signal from the VNA reaches the DUT. This is due primarily to leakage through the test set and reflections due to imperfect adapters. This creates measurement distortion due to the vector sum of the directivity signal combining with the reflected signal from the unknown DUT.

Frequency Tracking Error. Frequency-response tracking error, sometimes referred to as *reflection tracking error*, is when there is a variation in magnitude and phase versus frequency between the test and reference signal paths. This

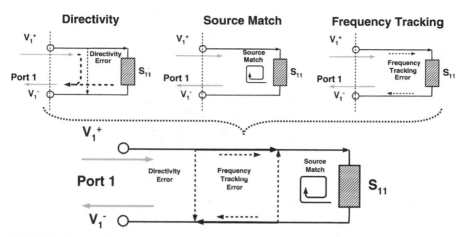

FIGURE 11.31 VNA one-port error model with the three error constants are computed during calibration to deliver accurate one port measurements up to measurement plane. (Adapted from the *HP 8510 VNA User's Manual.*)

is due to differences in length and loss between incident and test signal paths and imperfectly matched samplers.

Source Match Error. Source match error is due to a reflection between the source and DUT. Since the VNA source impedance is never exactly the same as the load being tested, some of the reflected signal is bounced back down the line, adding with the incident signal. This causes the incident signal magnitude and phase to vary as a function of frequency.

Two-Port Errors. Two-port measurements compute the transmission coefficients of a two-port device by taking the ratio of incident signal and transmitted signal. Two-port measurements S_{12} and S_{21} are often referred to as forward and reverse transmission. Sources of error are similar to the one-port described previously, with the addition of isolation and load match. Load match is similar to source match error except under two-port conditions. This refers to the impedance match between the transmission port and DUT. Reflections due to the mismatch between DUT and port 2 may be re-reflected back to port 1, which can cause error in S_{21}. Isolation error is due to the incident signal arriving at the receiver without actually passing through the DUT through an alternative path.

11.8.6. Calibration for One-Port Measurements

Extracting the previously described error constants versus frequency can eliminate these sources of VNA error. One-port error constants are calculated by measuring open, short, and load at the measurement plane (probe tip). This

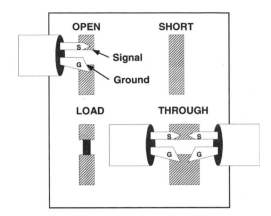

FIGURE 11.32 Typical microprobe calibration substrate with the open, load, short and thru terminations.

is completed using a calibration substrate that is matched to the probe type. The calibration substrate is made up of different structures that are designed to interface with the probe tip being used and have highly accurate electrical characteristics. The substrate is used to provide the accurate open, short, and load needed for extremely accurate measurements. The load is usually a laser-trimmed resistive component that is equal to the system impedance desired (usually 50 Ω). Many substrates include documentation that provides the electrical values to be entered into the VNA for each type of termination. This is done so that when the calibration is performed, the VNA can calculate the proper error constants. An example of a calibration substrate is shown in Figure 11.32 along with the various different load types.

It should be noted that the VNA source match error could still be an issue for cases when the calibration impedance is different from the DUT impedance being measured. Under these circumstances, custom-made substrates to calibrate the impedance of the VNA to the nominal impedance of the DUT can be manufactured.

11.8.7. Calibration for Two-Port Measurements

Two-port calibrations follow a process similar to that for the one-port, with the addition of one or more through-line test structures (a transmission line of the desired impedance) used to calibrate the S_{12} and S_{21} transmission coefficients.

11.8.8. Calibration Verification

After the calibration procedures have been completed, it is very important to ensure that calibration has been performed correctly. To verify the one-port calibration, the user should measure the substrate and compare the values across the frequency range. With the VNA format in Smith chart format, the user can read the values of the open, short, and load. The values should be as expected for the given measurement setup (the load should be at 1, the open should be at the right-hand side, and the short should be at the left-hand side of the Smith chart, as depicted in Figure 11.27). The two-port verification on the through test structures should be completed as well, to verify that the noise floor and propagation delay measurements match the expected value. Once the calibration has been verified to within the accuracy required, the measurements can begin.

Alternative Characteristic Impedance Formulas

These formulas should be used only when a field simulator is not available. A field simulator is required for the most accurate results.

A.1. MICROSTRIP

Refer to Figure A.1 when using the following formulas [Collins, 1992], which are accurate for $0.25 \leq W/H \leq 6$ and $1 < \varepsilon_r \leq 16$.

$$Z_o = \sqrt{\frac{\mu_0 \varepsilon_0}{\varepsilon_e}} \frac{1}{C_a}$$

$$C_a = \begin{cases} \dfrac{2\pi\varepsilon_0}{\ln(8H/W + W/4H)} & \text{when } \dfrac{W}{H} \leq 1 \\[3mm] \varepsilon_0 \left[\dfrac{W}{H} + 1.393 + 0.667\ln\left(\dfrac{W}{H} + 1.444\right) \right] & \text{when } \dfrac{W}{H} > 1 \end{cases}$$

$$\varepsilon_e = \frac{\varepsilon_r + 1}{2} + \frac{\varepsilon_r - 1}{2}\left(1 + \frac{12H}{W}\right)^{-1/2} + F - 0.217(\varepsilon_r - 1)\frac{T}{\sqrt{WH}}$$

$$F = \begin{cases} 0.02(\varepsilon_r - 1)\left(1 - \dfrac{W}{H}\right)^2 & \text{when } \dfrac{W}{H} < 1 \\[3mm] 0 & \text{when } \dfrac{W}{H} > 1 \end{cases}$$

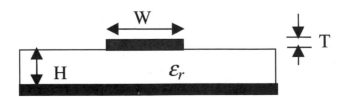

FIGURE A.1 Dimensions for use with the microstrip impedance formula.

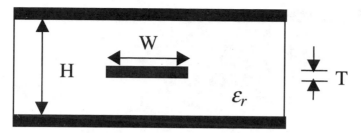

FIGURE A.2   Dimensions for use with the symmetrical stripline impedance formula.

A.2. SYMMETRIC STRIPLINE

Refer to Figure A.2 when using the following formulas [Johnson and Graham, 1993], which are accurate for $T/H < 0.25$ and $T/W < 0.11$.
    For $W/H < 0.35$,

$$Z_{0_{\text{sym, narrow}}} = \frac{60}{\sqrt{\varepsilon_r}} \ln \frac{4H}{\pi K_1}$$

$$K_1 = \left(\frac{W}{2}\right)\left[1 + \frac{T}{W\pi}\left(1 + \ln\frac{4\pi W}{T}\right) + 0.255\left(\frac{T}{W}\right)^2\right]$$

For $W/H > 0.35$,

$$Z_{0_{\text{sym, wide}}} = \frac{94.15}{\sqrt{\varepsilon_r}\left(\dfrac{W}{H-T} + \dfrac{K_2}{\pi}\right)}$$

$$K_2 = \frac{2}{1-T/H}\ln\left(\frac{1}{1-T/H} + 1\right)$$

$$-\left(\frac{1}{1-T/H} - 1\right)\ln\left(\frac{1}{(1-T/H)^2} - 1\right)$$

A.3. OFFSET STRIPLINE

Refer to Figure A.3 when using the following formula.
    The impedance for an offset stripline is calculated from the results of the symmetrical stripline formulas. The reader should note that this formula is an approximation and the accuracy of the results should be treated as

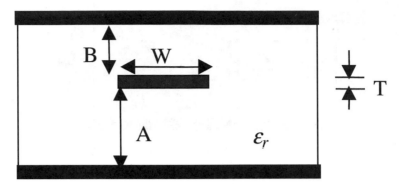

FIGURE A.3   Dimensions for use with the offset stripline impedance formula.

such. For more accurate results, use a field solver [Johnson and Graham, 1993].

$$Z_{o_{\text{offset}}} = 2\frac{Z_{o_{\text{sym}}}(H_1,W,T,\varepsilon_r)Z_{o_{\text{sym}}}(H_2,W,T,\varepsilon_r)}{Z_{o_{\text{sym}}}(H_1,W,T,\varepsilon_r) + Z_{o_{\text{sym}}}(H_2,W,T,\varepsilon_r)}$$

where $H_1 = 2A + T$ and $H_2 = 2B + T$.

GTL Current-Mode Analysis

GTL (Gunning Tranceiver Logic) buses are used widely in the industry for personal computer applications—specifically, the front-side bus, which is the interface between the processor and the chipset. In this appendix we provide the concepts required to analyze a GTL bus. For the most part, in this book we have concentrated on voltage mode analysis. In this appendix we step the reader though the analysis techniques used to predict the behavior of a GTL bus, and in parallel, teach the fundamentals of current-mode transmission line analysis.

The operation of a GTL bus is simple. The driver is simply an NMOS shunted to ground, and the other end of the transmission line is pulled up to the termination voltage. Turning the NMOS on and shunting the net to ground generates a low signal. The NMOS typically has a very low equivalent resistance. Turning the NMOS device off and letting the termination resistor pull the net high generates a high transaction.

B.1. BASIC GTL OPERATION

Consider Figure B.1a, which is an equivalent circuit for a GTL buffer (represented as a linear resistor and a switch) driving a transmission line in the steady-state low position. The GTL output buffer, which is typically just an NMOS device, is modeled as a linear resistor and a switch. In the steady-state low position, the bus will draw a dc current as calculated in equation (B.1) and the state low voltage will be calculated by equation (B.2):

$$I_L = \frac{V_{tt}}{R_{tt} + R(n)} \tag{B.1}$$

$$V_L = \frac{V_{tt}R(n)}{R_{tt} + R(n)} \tag{B.2}$$

When the switch opens (the NMOS turns off) and the net is pulled high, the transmission line will charge up. Subsequently, a current of $-I_L$ will be

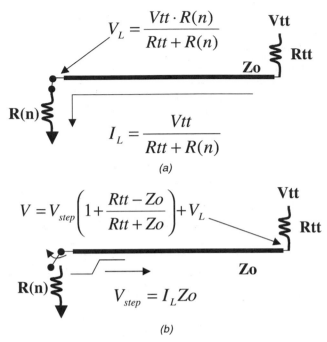

FIGURE B.1 GTL low-to-high bus transactions. (a) Steady-state low; (b) switch opens (NMOS turns off).

induced on the line and a voltage of V_{step},

$$V_{\text{step}} = I_L Z_o \tag{B.3}$$

will propagate toward the termination resistor R_{tt}. This is shown in Figure B.1b. The voltage at the receiver (it is assumed that the receiver is at the termination resistor) will depend on the reflection coefficient between the transmission line and the termination resistor:

$$V = V_{\text{step}} \left(1 + \frac{R_{tt} - Z_o}{R_{tt} + Z_o} \right) + V_L \tag{B.4}$$

In the steady-state high position, the net will assume a value of V_{tt}, as shown in Figure B.2. When the switch closes (the NMOS turns on), the voltage transmitted down the line will depend on V_{tt} and the voltage divider between the effective NMOS resistance and the transmission line:

$$V_{\text{step}} = -V_{tt} \frac{Z_o}{Z_o + R(n)} \tag{B.5}$$

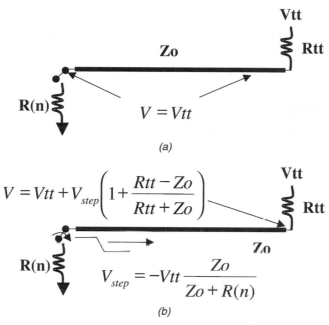

FIGURE B.2 GTL high-to-low bus transactions: (*a*) steady-state high; (*b*) switch closes.

Again, the voltage seen at the receiver will (it is assumed that the receiver is at the termination resistor) depend on the reflection coefficient between the transmission line and the termination resistor:

$$V = V_{tt} + V_{step}\left(1 + \frac{R_{tt} - Z_o}{R_{tt} + Z_o}\right) \tag{B.6}$$

Figure B.3 shows a generic example of a GTL bus data transition.

B.2. GTL TRANSITIONS WHEN A MIDDLE AGENT IS DRIVING

If additional agents are present on the bus, such as in the case of a dual-processor front-side-bus design, the voltages at the different agents can be calculated using the techniques detailed in Chapter 2. Although the reader should now have all the tools necessary to analyze a GTL type of bus, several useful examples and equations sets are presented.

If a three-load system (such as a dual-processor front-side bus with two processors and one chipset) is being driven by the middle agent, as shown in Figure B.4, the equations presented below apply. The GTL buffer is approximated by a switch and resistor shunted to ground, as depicted in Figure B.4.

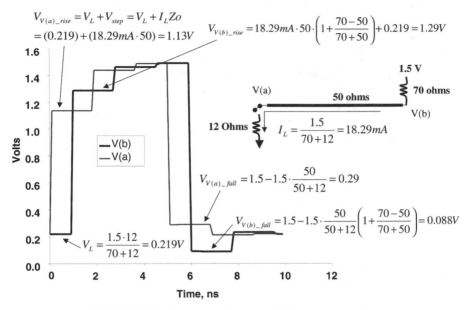

$$V_{V(a)\_rise} = V_L + V_{step} = V_L + I_L Zo$$
$$= (0.219) + (18.29mA \cdot 50) = 1.13V$$

$$V_{V(b)\_rise} = 18.29mA \cdot 50 \cdot \left(1 + \frac{70-50}{70+50}\right) + 0.219 = 1.29V$$

$$I_L = \frac{1.5}{70+12} = 18.29mA$$

$$V_{V(a)\_fall} = 1.5 - 1.5 \cdot \frac{50}{50+12} = 0.29$$

$$V_{V(b)\_fall} = 1.5 - 1.5 \cdot \frac{50}{50+12}\left(1 + \frac{70-50}{70+50}\right) = 0.088V$$

$$V_L = \frac{1.5 \cdot 12}{70+12} = 0.219V$$

FIGURE B.3   Example of a basic GTL bus transaction.

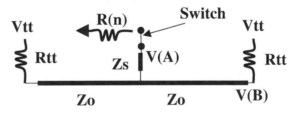

FIGURE B.4   Special case: three-load GTL bus driving from the middle agent.

When the buffer switches from a *low to a high state*, it is necessary to calculate the conditions on the bus just prior to the transaction (when the NMOS turns off). If it is assumed that the bus was previously in the steady-state low condition, the equations

$$I_L = \frac{V_{tt}}{\frac{1}{2}R_{tt} + R(n)} \tag{B.7}$$

$$V_L = \frac{V_{tt}R(n)}{\frac{1}{2}R_{tt} + R(n)} \tag{B.8}$$

can be used to determine the steady-state low current and voltage. When the switch opens, the initial voltage step injected onto the transmission line is calculated as

$$V_{delta} = I_L Z_s \tag{B.9}$$

The signal will see a reflection coefficient at the junction of the stub and the main bus as calculated using equation (B.10) and the transmission coefficient is calculated in (B.11).

$$\rho_{at\ stub} = \frac{Z_o \| Z_o - Z_s}{Z_o \| Z_o + Z_s} \tag{B.10}$$

$$T = 1 + \rho_{at\ stub} \tag{B.11}$$

The initial voltage seen on the bus at the driver is the initial voltage step plus the dc offset caused by the voltage divider between the termination pull-up resistor and the equivalent NMOS resistance:

$$V_{initial} = V_{delta} + V_L \tag{B.12}$$

Finally, the voltage due to the first reflection seen at the receiver $V(B)$ is calculated:

$$V(B) = TV_{delta}\left(1 + \frac{R_{tt} - Z_o}{R_{tt} + Z_o}\right) + V_L \tag{B.13}$$

When the buffer switches from a *high to a low state* (the switch closes, or the NMOS turns on), the voltage step injected on the line will depend on V_{tt} and the voltage divider between the stub impedance and the NMOS equivalent resistance:

$$V_{delta} = -V_{tt}\frac{Z_s}{Z_s + R(n)} \tag{B.14}$$

Subsequently, the initial voltage seen on the bus is V_{tt} plus equation (B.14) (which is negative):

$$V_{initial} = V_{tt} + V_{delta} \tag{B.15}$$

The voltage seen at the receiver, $V(B)$, depends on the transmission coefficient at the stub junction:

$$V(B) = V_{tt} + TV_{delta}\left(1 + \frac{R_{tt} - Z_o}{R_{tt} + Z_o}\right) \tag{B.16}$$

B.3. GTL TRANSITIONS WHEN AN END AGENT WITH A TERMINATION IS DRIVING

Consider Figure B.5.

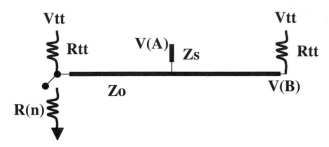

FIGURE B.5   Special case: three-load GTL bus driving from the end agent.

Low-to-High Transition   The switch opens, simulating the NMOS turning off.

$$I_L = \frac{V_{tt}}{\frac{1}{2}R_{tt} + R(n)}$$

$$V_L = \frac{V_{tt}R(n)}{\frac{1}{2}R_{tt} + R(n)}$$

$$V_{\text{delta}} = I_L(Z_o \,\|\, R_{tt}) = \frac{I_L Z_o R_{tt}}{Z_o + R_{tt}}$$

$$\rho_{\text{at stub}} = \frac{Z_o \,\|\, Z_s - Z_o}{Z_o \,\|\, Z_s + Z_o}$$

$$T = 1 + \rho_{\text{at stub}}$$

$$V_{\text{initial}} = V_{\text{delta}} + V_L$$

$$V(A) = 2TV_{\text{delta}} + V_L$$

$$V(B) = TV_{\text{delta}}\left(1 + \frac{R_{tt} - Z_o}{R_{tt} + Z_o}\right) + V_L$$

High-to-Low Transition   The switch closes, simulating the NMOS turning on.

$$V_L = \frac{V_{tt}R(n)}{\frac{1}{2}R_{tt} + R(n)}$$

$$V_{\text{delta}} = -V_{tt}\frac{R_{tt} \,\|\, Z_o}{R_{tt} \,\|\, Z_o + R(n)}$$

$$\rho_{\text{at stub}} = \frac{Z_o \,\|\, Z_s - Z_o}{Z_o \,\|\, Z_s + Z_o}$$

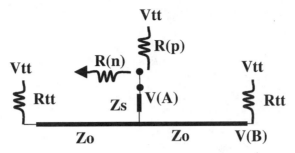

FIGURE B.6 Special case: three-load GTL bus with an extra pull up in the middle.

$$T = 1 + \rho_{\text{at stub}}$$

$$V_{\text{initial}} = V_{tt} + V_{\text{delta}}$$

$$V(A) = V_{tt} + 2TV_{\text{delta}}$$

$$V(B) = V_{tt} + TV_{\text{delta}} \left(1 + \frac{R_{tt} - Z_o}{R_{tt} + Z_o} \right)$$

B.4. TRANSITIONS WHEN THERE IS A PULL-UP AT THE MIDDLE AGENT

Consider Figure B.6.

Low-to-High Transition

$$I_L = \frac{V_{tt}}{\frac{1}{2}R_{tt} + R(n)}$$

$$V_L = \frac{V_{tt}R(n)}{\frac{1}{2}R_{tt} + R(n)}$$

$$I_L + \frac{V_{tt} - V_{\text{initial}}}{R(p)} + \frac{V_L - V_{\text{initial}}}{Z_s} = 0$$

$$V_{\text{initial}} = \frac{R(p)Z_s}{R(p) + Z_s} \left(I_L + \frac{V_{tt}}{R(p)} + \frac{V_L}{Z_s} \right)$$

$$\rho_{\text{at stub}} = \frac{Z_o \| Z_o - Z_s}{Z_o \| Z_o + Z_s}$$

$$T = 1 + \rho_{\text{at stub}}$$

$$V(B) = T(V_{\text{initial}} - V_L) \left(1 + \frac{R_{tt} - Z_o}{R_{tt} + Z_o} \right) + V_L$$

High-to-Low Transition

$$V_L = \frac{V_{tt}R(n)}{\frac{1}{2}R_{tt} + R(n)}$$

$$V_{\text{delta}} = -V_{tt}\frac{Z_s}{Z_s + R(n)}$$

$$\rho_{\text{at stub}} = \frac{Z_o \| Z_o - Z_s}{Z_o \| Z_o + Z_s}$$

$$T = 1 + \rho_{\text{at stub}}$$

$$V_{\text{initial}} = V_{tt} + V_{\text{delta}}$$

$$V(B) = V_{tt} + TV_{\text{delta}}\left(1 + \frac{R_{tt} - Z_o}{R_{tt} + Z_o}\right)$$

Frequency-Domain Components in a Digital Signal

Throughout this book, several frequency-dependent effects have been discussed that have a significant effect on the signal integrity of a digital signal propagating on a bus. For example, skin effect resistance, inductance, and the dielectric constant all have values that change with frequency. The problem with analyzing these effects is that a digital signal is not sinusoidal; thus the signal is composed of an infinite number of sinusoidal components that, when summed, produce the familiar digital signal shape. Since digital signals are approximated as square (or trapezoidal) waves, the spectral frequency components can be calculated as

$$f(t) = \frac{2}{\pi} \sum_{n-1,3,5...} \frac{1}{n} \sin 2\pi nFt \qquad (C.1)$$

which is the Fourier expansion of a periodic square wave with a 50% duty cycle, where F is the frequency and t is time. A 100-MHz periodic square wave, for example, is a superposition of an infinite number of sinusoidal functions with frequencies that are odd multiples of the fundamental (i.e., 100 MHz, 300 MHz, 500 MHz, etc.). These components are referred to as the *harmonics*, where $n = 1$ refers to the first, or fundamental, harmonic (100 MHz); $n = 3$ refers to the third harmonic (300 MHz); and so on. As the frequency components get higher, the sinusoidal functions become smaller in magnitude. Subsequently, the peaks of the frequency components will form a spectral envelope, as depicted in Figure 10.3. It should be noted that in reality, a digital waveform would also exhibit even harmonics because the signal is rarely a perfect 50% duty cycle, even for clocking signals.

During the analysis of a bus, it is often necessary to approximate the frequency components of a digital signal. There are several ways to do this. For example, if it is possible to make laboratory measurements of a system, the frequency components can be measured directly with a spectrum analyzer. Alternatively, waveforms from simulations or measurements can be analyzed with the fast Fourier transform (FFT), which is available in a wide variety of tools, including Microsoft Exel, Mathematica, and Mathcad. The FFT will output a spectrum of all the frequency harmonics that compose a waveform. Often, the FFT is useful for including frequency-dependent components into the simulation by transforming the time-domain signal into the frequency domain, manipulating the data and performing an inverse FFT back into the time domain (such as including frequency-dependent attenuation effects, as shown in Figure 4.10).

There are several ways to estimate the spectral envelope of a digital signal. Some engineers approach this problem by assuming that the frequencies above the fifth harmonic will be negligible and thus estimate that the highest spectral component will be at five times the fundamental. For example, the fifth harmonic of a periodic 100-MHz digital waveform will be 500 MHz. Another, more widely used estimation technique approximates the frequency envelope by observing the fastest edge rate that will be used in the system. The approximation is derived here.

The standard method of measuring the rise or fall times of the signal is to look at the time it takes for the signal to rise from 10% to 90% of the maximum signal swing. The step response of a signal (a rising edge with an infinitely fast rise time) into a network with a time constant τ is governed by the equation

$$V = V_{input}(1 - e^{-t/\tau})$$ (C.2)

where t is time and V_{input} is the input voltage. If V_{input} is assumed to be 1 V, the time it takes for the voltage V to rise to 0.9 V, or 90% of the signal swing, is calculated as

$$t_{90\%} = 2.3\tau$$ (C.3)

The signal swing to 0.1 V, or 10% of the maximum swing, is

$$t_{10\%} = 0.105\tau$$ (C.4)

Subsequently, the time it takes a step to transition from 10% to 90% of the signal swing is

$$t_{10-90\%} = (2.3 - 0.105)\tau = 2.195\tau$$ (C.5)

The frequency response of a network is with a time constant of τ (i.e., RC)

$$F_{3\ dB} = \frac{1}{2\pi\tau} \rightarrow \tau = \frac{1}{2\pi F_{3\ dB}}$$ (C.6)

If (C.6) is inserted into (C.5), the result is

$$t_{10-90\%} = \frac{1.09}{\pi F_{3\ dB}} = \frac{0.35}{F_{3\ dB}} \approx \frac{1}{\pi F_{3\ dB}} \rightarrow F_{3\ dB} = \frac{0.35}{t_{10-90\%}}$$ (C.7)

Equation (C.7) is the resulting 10 to 90% rise or fall time of a perfect step function driven through a network with a time constant of τ. Obviously, this does not apply to real systems because realistic edge rates are finite. However, the $F_{3\,dB}$ frequency is still a reasonable approximation of the spectral envelope of a digital pulse because below this frequency is where most of the spectral energy is contained (as depicted in Figure 10.3).

When a signal passes through a network with a time constant of τ, the edge rate will be degraded because of the filtering effects. The degradation of a finite edge rate can be approximated by the equation

$$t \approx \sqrt{t_{input}^2 + t_{step}^2} \qquad (C.8)$$

where t_{input} is the input rise or fall time and t_{step} is the result from equation (C.7).

The reader should note that the relationships above between the edge rate and the spectral content are only approximations, however, they tend to be extremely useful for back-of-the-envelope calculations and for determining the capability of laboratory equipment. For example, an oscilloscope with an input bandwidth of 3 GHz will be able to measure edge rates correctly only when they are greater than approximately $0.35/3 \times 10^9 \approx 117$ ps. If it were attempted to measure a 50-ps rise time on such a scope, for example, the waveform on the screen would have an edge rate of approximately $\sqrt{50 \text{ ps}^2 + 117 \text{ ps}^2} = 127$ ps.

Useful *S*-Parameter Conversions

D.1. *ABCD*, *Z*, AND *Y* PARAMETERS

ABCD and *Z* parameters are defined in this appendix, however, they are not derived, and a minimal explanation is presented. If the reader wishes to understand the nature of these conversions, any microwave systems textbook may be consulted. Although the full explanation of these relationships is beyond the scope of this book, they are presented here because thy can be extremely useful for extracting electrical parameters from vector network analyzer (VNA) measurements.

The equations that follow convert *S*-parameters to *ABCD*, *Y*, and *Z* parameters [Pozar, 1990]. Figure D.1 depicts how *ABCD* parameters relate to specific

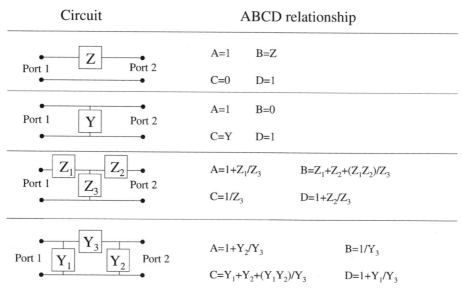

Circuit	ABCD relationship

Z — Port 1 / Port 2	$A=1$ $B=Z$ $C=0$ $D=1$
Y — Port 1 / Port 2	$A=1$ $B=0$ $C=Y$ $D=1$
Z_1 Z_2 Z_3 — Port 1 / Port 2	$A=1+Z_1/Z_3$ $\quad$ $B=Z_1+Z_2+(Z_1Z_2)/Z_3$ $C=1/Z_3$ $\quad$ $D=1+Z_2/Z_3$
Y_3 Y_1 Y_2 — Port 1 / Port 2	$A=1+Y_2/Y_3$ $\quad$ $B=1/Y_3$ $C=Y_1+Y_2+(Y_1Y_2)/Y_3$ $\quad$ $D=1+Y_1/Y_3$

FIGURE D.1 Relationship between some basic circuit configurations and the *ABCD* parameters.

332

circuit topologies [Pozar, 1990]. Note that in these equations, the impedance Z_o is the system impedance to which the VNA was calibrated.

$$S_{11} = \frac{A + B/Z_o - CZ_o - D}{A + B/Z_o + CZ_o + D}$$

$$S_{12} = \frac{2(AD - BC)}{A + B/Z_o + CZ_o + D}$$

$$S_{21} = \frac{2}{A + B/Z_o + CZ_o + D}$$

$$S_{22} = \frac{-A + B/Z_o - CZ_o + D}{A + B/Z_o + CZ_o + D}$$

Conversely, *ABCD* parameters can be converted to *S* parameters as follows:

$$A = \frac{(1 + S_{11})(1 - S_{22}) + S_{12}S_{21}}{2S_{21}}$$

$$B = Z_o\frac{(1 + S_{11})(1 + S_{22}) - S_{12}S_{21}}{2S_{21}}$$

$$C = \frac{(1 - S_{11})(1 - S_{22}) - S_{12}S_{21}}{2Z_oS_{21}}$$

$$D = \frac{(1 - S_{11})(1 + S_{22}) - S_{12}S_{21}}{2S_{21}}$$

The relationships between the *Z* parameters (impedance matrix) and the *S* parameters are as follows:

$$Z_{11} = Z_o\frac{(1 + S_{11})(1 - S_{22}) + S_{12}S_{21}}{(1 - S_{11})(1 - S_{22}) - S_{12}S_{21}}$$

$$Z_{12} = Z_o\frac{2S_{12}}{(1 - S_{11})(1 - S_{22}) - S_{12}S_{21}}$$

$$Z_{21} = Z_o\frac{2S_{21}}{(1 - S_{11})(1 - S_{22}) - S_{12}S_{21}}$$

$$Z_{22} = Z_o\frac{(1 - S_{11})(1 + S_{22}) + S_{12}S_{21}}{(1 - S_{11})(1 - S_{22}) - S_{12}S_{21}}$$

The conversions from S to Z parameters are as follows:

$$S_{11} = \frac{(Z_{11} - Z_o)(Z_{22} + Z_o) - Z_{12}Z_{21}}{(Z_{11} + Z_o)(Z_{22} + Z_o) - Z_{12}Z_{21}}$$

$$S_{12} = \frac{2Z_{12}Z_o}{(Z_{11} + Z_o)(Z_{22} + Z_o) - Z_{12}Z_{21}}$$

$$S_{21} = \frac{2Z_{21}Z_o}{(Z_{11} + Z_o)(Z_{22} + Z_o) - Z_{12}Z_{21}}$$

$$S_{22} = \frac{(Z_{11} + Z_o)(Z_{22} - Z_o) - Z_{12}Z_{21}}{(Z_{11} + Z_o)(Z_{22} + Z_o) - Z_{12}Z_{21}}$$

The conversions between the S and Y parameters are as follows:

$$Y_o = \frac{1}{Z_o}$$

$$S_{11} = \frac{(Y_o - Y_{11})(Y_{11} + Y_o) + Y_{21}Y_{12}}{(Y_{11} + Y_o)(Y_{22} + Y_o) - Y_{12}Y_{21}}$$

$$S_{12} = \frac{-2Y_{12}Y_o}{(Y_{11} + Y_o)(Y_{22} + Y_o) - Y_{12}Y_{21}}$$

$$S_{21} = \frac{-2Y_{21}Y_o}{(Y_{11} + Y_o)(Y_{22} + Y_o) - Y_{12}Y_{21}}$$

$$S_{22} = \frac{(Y_o + Y_{11})(Y_o - Y_{22}) + Y_{21}Y_{12}}{(Y_{11} + Y_o)(Y_{22} + Y_o) - Y_{12}Y_{21}}$$

The conversions between the Y and S parameters are as follows:

$$Y_{11} = Y_o \frac{(1 - S_{11})(1 + S_{22}) + S_{12}S_{21}}{(1 + S_{11})(1 + S_{22}) - S_{12}S_{21}}$$

$$Y_{12} = Y_o \frac{-2S_{12}}{(1 + S_{11})(1 + S_{22}) - S_{12}S_{21}}$$

$$Y_{21} = Y_o \frac{-2S_{21}}{(1 + S_{11})(1 + S_{22}) - S_{12}S_{21}}$$

$$Y_{22} = Y_o \frac{(1 + S_{11})(1 - S_{22}) + S_{12}S_{21}}{(1 + S_{11})(1 + S_{22}) - S_{12}S_{21}}$$

D.2. NORMALIZING THE *S* MATRIX TO A DIFFERENT CHARACTERISTIC IMPEDANCE

As described in Section 11.8, it is sometimes necessary to normalize the *S*-parameters to an impedance other than that used for the measurement. For example, if the attenuation of a line must be measured, it is necessary to calibrate the VNA to the nominal impedance of the transmission line. As described in Chapter 11, this is done by calibrating with a calibration substrate that is equal (or very close to) the nominal impedance of the transmission line. If this is not done, accurate results cannot be obtained because reflections on the line will interfere with proper measurement. If the characteristic impedance of the transmission line is measured (as described in Chapter 11), the *S* parameters can be manipulated mathematically so that will appear as if they were measured with a VNA impedance that matches the characteristic impedance of the transmission line (i.e., there will be no reflections present).

The procedure is as follows. Initially, the characteristic impedance of the transmission line under test, $Z_{o_{DUT}}$, must be determined as described in Chapter 11. The *S*-parameter matrix (the scattering matrix), S_m, is measured with a VNA as a function of frequency, which is calibrated to a system impedance of Z_{o_1}.

$$S_m = \begin{vmatrix} S_{11} & S_{12} \\ S_{21} & S_{22} \end{vmatrix}$$

The S_m matrix is normalized to $Z_{o_{DUT}}$ by converting the *S*-parameter matrix, S_m, to the *Z*-parameter matrix, Z_m, using the conversion factors presented in Section D.1 with $Z_o = Z_{o_1}$.

$$Z_m = \begin{vmatrix} Z_{11} & Z_{12} \\ Z_{21} & Z_{22} \end{vmatrix}$$

Finally, the Z_m matrix is converted back to *S* parameters using the conversions in Section D.1 with $Z_o = Z_{o_{DUT}}$. This produces the normalized matrix, S'_m. To verify that S'_m was correctly normalized to the characteristic impedance of the transmission line, verify that the components S_{11} and S_{22} are very small.

$$S'_m = \begin{vmatrix} S'_{11} & S'_{12} \\ S'_{21} & S'_{22} \end{vmatrix} \approx \begin{vmatrix} 0 & S'_{12} \\ S'_{21} & 0 \end{vmatrix} = \text{normalized } S \text{ matrix}$$

The normalized *S* matrix can be used with equation (11.23) to extract out the attenuation of a transmission line.

D.3. DERIVATION OF THE FORMULAS USED TO EXTRACT THE MUTUAL INDUCTANCE AND CAPACITANCE FROM A SHORT STRUCTURE USING S_{21} MEASUREMENTS

Refer to Figure 11.29. The S_{21} measurement of the open structure is used to extract the mutual capacitance. Since the current in the inductors will be negligible, the primary coupling mechanism between ports 1 and 2 is the mutual capacitance, which will look like a series impedance similar to the top circuit in Figure D.1. The following is a derivation of equation (11.16):

$$S_{21} = \frac{2}{A + B/Z_o + CZ_o + D}$$

Applying the conditions of Figure D.1 ($A = 1$, $B = Z$, $C = 0$, and $D = 1$) yields the following:

$$S_{21} = \frac{2}{1 + Z/Z_o + 1} = \frac{2}{2 + Z/Z_o}$$

Substituting the impedance of the mutual capacitance ($1/j\omega C_{12}$) for Z and simplifying yields the following, which is the same as equation (11.16):

$$S_{21} = \frac{2Z_o j\omega C_{12}}{2 j\omega C_{12} Z_o + 1}$$

The S_{21} measurement of the shorted structure in Figure 11.29 is used to extract the mutual inductance. In this case there will be negligible voltage across the mutual capacitor since the two legs are both shorted to ground. Subsequently, the primary coupling mechanism between ports 1 and 2 is the mutual inductance. The following is a derivation of equation (11.18).

To solve this system, the voltage equations must be written. V_1 refers to the voltage at port 1 and V_2 refers to the voltage at port 2. It is also assumed that $L_1 = L_2 = L$.

$$V_1 = j\omega L I_1 + j\omega L_M I_2$$
$$V_2 = j\omega L I_2 + j\omega L_M I_1$$

This yields the following matrix equation:

$$\begin{vmatrix} V_1 \\ V_2 \end{vmatrix} = \begin{vmatrix} j\omega L & j\omega L_M \\ j\omega L_M & j\omega L \end{vmatrix} \cdot \begin{vmatrix} I_1 \\ I_2 \end{vmatrix}$$

The impedance matrix (Z parameters) is in the form of the following:

$$\begin{vmatrix} V_1 \\ V_2 \end{vmatrix} = \begin{vmatrix} Z_{11} & Z_{12} \\ Z_{21} & Z_{22} \end{vmatrix} \cdot \begin{vmatrix} I_1 \\ I_2 \end{vmatrix}$$

Subsequently, S_{21} can be related to the Z parameters as described in Section D.1.

$$S_{21} = \frac{2Z_{21}Z_o}{(Z_{11} + Z_o)(Z_{22} + Z_o) - Z_{21}Z_{12}}$$

$$= \frac{2(j\omega L_M)Z_o}{(j\omega L + Z_o)(j\omega L + Z_o) - (j\omega L_M)(j\omega L_M)}$$

$$S_{21} = \frac{2(j\omega L_M)Z_o}{\omega^2 L_M^2 - \omega^2 L^2 + 2j\omega L Z_o + Z_o^2} \approx \frac{2(j\omega L_M)Z_o}{2j\omega L Z_o + Z_o^2}$$

The approximation is made in the last step because the term $\omega^2 L_M^2 - \omega^2 L^2$ is usually very small. Simplification yields the following equation, which is the same as equation (11.18):

$$S_{21} \approx \frac{2j\omega L_M}{2j\omega L + Z_o}$$

Alternatively, if $2j\omega L \ll Z_o$, then the equation reduces to the following:

$$S_{21} \approx \frac{2j\omega L_M}{Z_o}$$

D.4. DERIVATION OF THE FORMULA TO EXTRACT SKIN EFFECT RESISTANCE FROM A TRANSMISSION LINE

If the losses due to the dielectric are small compared to the skin effect resistance, and the VNA calibrated impedance is close to the impedance of the transmission line being measured, S_{21} can be related to the skin effect resistance (ac resistance) as follows. As per Figure D.1 for a series impedance,

$$ABCD = \begin{vmatrix} A & B \\ C & D \end{vmatrix} = \begin{vmatrix} 1 & Z \\ 0 & 1 \end{vmatrix}$$

Subsequently, S_{21} can be related to Z, which is the ac resistance.

$$S_{21} = \frac{2}{A + B/Z_o + CZ_o + D} = \frac{2}{1 + R_{ac}/Z_o + (0)Z_o + 1} - \frac{2Z_o}{2Z_o + R_{ac}}$$

where Z_o is the VNA system impedance and R_{ac} is the skin effect resistance.

Definition of the Decibel

The unit of decibels is used heavily in both EMI analysis and frequency-domain measurements using a vector network analyzer (VNA). Subsequently, it is necessary to define the term so that the reader can accurately interpret data presented with units of decibels.

The *bel*, as defined in equation (E.1), is the basic unit from which the decibel is derived:

$$\text{bel} = \log_{10} \frac{P_2}{P_1} \tag{E.1}$$

where P_1 is the measured reference power and P_2 is the power present at the measurement.

Because the bel results in coarse measurement, a more useful metric is achieved with the *decibel* (dB):

$$\text{dB} = 10 \log_{10} \frac{P_2}{P_1} \tag{E.2}$$

Another form of the decibel is

$$\text{dB} = 10 \log_{10} \frac{P_2}{P_1} = 10 \log_{10} \frac{V_2^2 / R_2}{V_1^2 / R_1} \tag{E.3}$$

Assuming that $R_1 = R_2$, which is usually the case, the decibel takes the form

$$\text{dB} = 10 \log_{10} \frac{V_2^2 / R_2}{V_1^2 / R_1} = 10 \log_{10} \frac{V_2^2}{V_1^2} = 20 \log_{10} \frac{V_2}{V_1} \tag{E.4}$$

Alternatively, the decibel can be expressed in terms of current assuming that $R_1 = R_2$:

$$\text{dB} = 10 \log_{10} \frac{I_2^2 R_2}{I_1^2 R_1} = 10 \log_{10} \frac{I_2^2}{I_1^2} = 20 \log_{10} \frac{I_2}{I_1} \tag{E.5}$$

For the purposes of describing attenuation, the usually decibel takes the form $20 \log(M)$, where M is the magnitude of the quantity being described. For example, if a radiated emission were attenuated by a factor of 3, the resulting attenuation, as measured in decibels, would be $20\log(\frac{1}{3}) = -9.54$ dB. The reader should note that the negative sign is often neglected; subsequently, this example would be reported as 9.54 dB of loss, or the signal has been attenuated by 9.54 dB. A signal that experiences a gain of 3 would be calculated in decibels as $20\log(3) = 9.54$ dB. Again, the gain in this example would be reported as a gain of 9.54 dB.

FCC Emission Limits

Table F.1 shows the FCC emission limits for class A and class B digital devices. A class B digital device is one marketed for use in a residential environment. A class A digital device is one marketed for use in a commercial, industrial, or business environment. Note that the specified measurement distances are 10 m for class A and 3 m for class B. Thus class A emission limitations are much more relaxed. Figure F.1 shows the limits graphically. Figure F.1 also shows the open-box emission limitations for a class B device. The open-box limitations apply to computer motherboards that are marketed as a component rather then a complete system and are measured with the chassis cover removed.

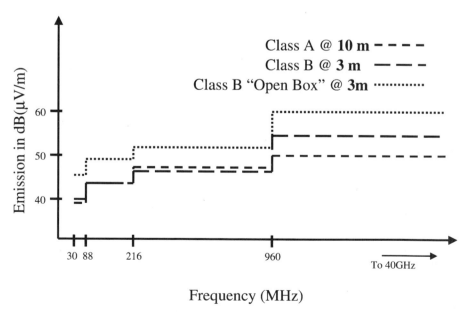

FIGURE F.1 FCC emission limitations and measurement distances for digital devices.

TABLE F.1. FCC Emission Limits for Digital Devices

	Emission Frequency	Field Strength	
Class A (at 10 m)	30 to 88 MHz	90 μV/m	39.08 dB(μV/m)
	88 to 216 MHz	150 μV/m	43.52 dB(μV/m)
	216 to 960 MHz	210 μV/m	46.44 dB(μV/m)
	960 MHz to 40 GHz	300 μV/m	49.54 dB(μV/m)
Class B (at 3 m)	30 to 88 MHz	100 μV/m	40.00 dB(μV/m)
	88 to 216 MHz	150 μV/m	43.52 dB(μV/m)
	216 to 960 MHz	200 μV/m	46.02 dB(μV/m)
	960 MHz to 40 GHz	500 μV/m	53.98 dB(μV/m)

■■■■■ BIBLIOGRAPHY

Byers, Andrew, Melinda Piket-May, and Stephen Hall, Quantifying the impact of a non-ideal ground return path, *Proceedings of the International Conference and Exhibition on High Density Packaging and MCMs*, April 1999.

Cheng, David, *Field and Wave Electromagnetics*, Addison-Wesley, Reading, MA, 1989.

Collins, Robert, *Foundations for Microwave Engineering*, McGraw-Hill, New York, 1992.

Coombs, Clyde, *Printed Circuits Handbook*, 4th ed., McGraw-Hill, New York, 1996.

DeFalco, John, Reflection and crosstalk in logic circuit interconnections, *IEEE Spectrum*, 1970.

Hubing, Todd, et al., Power bus decoupling on multi-layer printed circuit boards, *IEEE Transactions on Electromagnetic Compatibility*, Vol. 37, No. 2, May 1995.

IPC-2141, *Controlled Impedance Circuit Boards and High Speed Logic Design*, Institute for Interconnecting and Packaging Electronic Circuits, April 1996.

Johnk, Carl, *Engineering Electromagnetic Fields and Waves*, 2nd ed., Wiley, 1988.

Johnson, Howard and Martin Graham, *High-Speed Digital Design: A Handbook of Black Magic*, Prentice Hall, Upper Saddle River, NJ, 1993.

Liaw, Jyh-Haw and Henri Merkelo, Signal integrity issues at a split ground and power planes, *Proceedings of the 46th Electronic Components and Technology Conference*, 1996.

Mardiguian, Michel, *Controlling Radiated Emissions by Design*, Van Nostrand Reinhold, New York, 1992.

McCall, James, Successful PCB testing methodology, *Printed Circuit Design*, June 1999.

Mumby, Steven, *Dielectric Properties of FR-4 Laminates as a Function of Thickness and the Electrical Frequency of Measurements* (IPC-TP-749), Institute for Interconnecting and Packaging Electronic Circuits, 1988.

Ott, Henry, *Noise Reduction Techniques in Electronic Systems*, 2nd ed., Wiley, New York, 1988.

Poon, Ron, *Computer Circuits Electrical Design*, Prentice Hall, Upper Saddle River, NJ, 1995.

Pozar, David, *Microwave Engineering*, Addison-Wesley, Reading, MA, 1990.

Ramo, Simon, John Whinnery, and Theodore Van Duzer, *Fields and Waves in Communications Electronics*, Wiley, New York, 1994.

Sedra, Adel and Kenneth Smith, *Microelectronic Circuits*, 3rd ed., Saunders College Publishing, Philadelphia, 1991.

Selby, Samuel, *Standard Mathematical Tables*, 21st ed., CRC Press, Boca Raton, FL, 1973.

White, Donald, *Shielding Design Methodology and Procedures*, Interference Control Technology, Inc., Gainesville, VA, 1986.

Zwillinger, Daniel, *Standard Mathematical Tables and Formulae*, 30th ed., CRC Press, Boca Raton, FL, 1996.